A PLUME BOOK

RELATIVITY

ROGER PENROSE is Emeritus Rouse Ball Professor of Mathematics at Oxford University. He is the author of a number of books, including *The Road to Reality: A Complete Guide to the Laws of the Universe* and *The Emperor's New Mind: Concerning Computers, Minds, and the Laws of Physics*.

ROBERT GEROCH is a professor of physics at the University of Chicago. He received his Ph.D. at Princeton University, and has held positions at the University of London, Syracuse University, and the University of Texas. He is the author of *General Relativity from A to B*.

DAVID C. CASSIDY is a professor of natural sciences at Hofstra University. He served as associate editor of *The Collected Papers of Albert Einstein,* and is the author of *Einstein and Our World, Uncertainty: The Life and Science of Werner Heisenberg,* and *J. Robert Oppenheimer and the American Century*.

RELATIVITY

The Special and General Theory

Albert Einstein

Translated by Robert W. Lawson

Introduction by Roger Penrose
Commentary by Robert Geroch
with a Historical Essay by David C. Cassidy

A PLUME BOOK

PLUME
Published by Penguin Group
Penguin Group (USA) Inc., 375 Hudson Street, New York, New York 10014, USA
Penguin Group (Canada), 90 Eglinton Avenue East, Suite 700, Toronto,
Ontario Canada M4P 2Y3 (a division of Pearson Penguin Canada Inc.)
Penguin Books Ltd., 80 Strand, London WC2R 0RL, England
Penguin Ireland, 25 St. Stephen's Green, Dublin 2,
Ireland (a division of Penguin Books Ltd.)
Penguin Group (Australia), 250 Camberwell Road, Camberwell, Victoria 3124,
Australia (a division of Pearson Australia Group Pty. Ltd.)
Penguin Books India Pvt. Ltd., 11 Community Centre, Panchsheel Park,
New Delhi – 110 017, India
Penguin Group (NZ), cnr Airborne and Rosedale Roads, Albany,
Auckland 1310, New Zealand (a division of Pearson New Zealand Ltd.)
Penguin Books (South Africa) (Pty.) Ltd., 24 Sturdee Avenue,
Rosebank, Johannesburg 2196, South Africa

Penguin Books Ltd., Registered Offices: 80 Strand, London WC2R 0RL, England

Published by Plume, a member of Penguin Group (USA) Inc. Previously published
in a Pi Press edition.

First Plume Printing, September 2006
10   9   8   7   6   5   4   3   2   1

Commentary and Historical Essay © 2005 by Pearson Education, Inc.

Introduction © by Roger Penrose 2004, 2005

Original version of Roger Penrose's Introduction first published by The Folio Society
Ltd in 2004. Reprinted by arrangement with The Folio Society Ltd.

 REGISTERED TRADEMARK—MARCA REGISTRADA

CIP data is available.
ISBN 0-13-186261-8 (hc.)
ISBN 0-452-28784-7 (pbk.)

Printed in the United States of America

CONTENTS

PART II
THE GENERAL THEORY OF RELATIVITY

Contents

INTRODUCTION

Roger Penrose

Albert Einstein put forward his special theory of relativity in 1905, and he fully formulated his general theory ten years later, in 1915, publishing it in 1916. His account of these theories in this book *Relativity: The Special and General Theory* first appeared (in the original German edition) also in 1916. It is remarkable—and, in my view, very gratifying—that he should have produced such an account for the lay public so soon after his general theory was completed. That theory, in particular, presented an extraordinary and revolutionary view of the physical world, providing profound new insights for our conceptions of space, time, and gravitation. This view has now superbly survived the tests of time. When Einstein first put forward his general theory almost ninety years ago, there was inevitably a deep and widespread curiosity, among the general public, as to the actual nature of his revolutionary ideas, and there was an equally inevitable scope for misunderstandings of various kinds. There was, accordingly, an immediate need for an accessible account from the master himself. *Relativity* helped enormously in minimizing such misunderstandings. But, even today, there are many physicists who have not fully come to terms with the revolutions in physical world-view that Einstein presented. The reprinting of Einstein's account to coincide with the centenary of the special theory is therefore indeed timely. In this edition we are also fortunate in having the advantage of an excellent exposition of Einstein's foundational ideas on relativity from a more modern perspective in Robert Geroch's commentary.

It should be pointed out that of these two relativity revolutions, it was the general theory that was Einstein's most uniquely original contribution. Special relativity—the relativity theory that Einstein put forward, among other fundamental advances, in what has become known as his "Miraculous Year" of 1905—was far less uniquely Einstein's own achievement. That theory was concerned with the puzzling transformations that mix up space and time, as is needed when the relative velocity between the reference frames of different observers approaches that of light. But these transformations had been in the air for some time prior to Einstein entering the scene. Others before him, particularly the outstanding Dutch physicist Hendrik

Antoon Lorentz and the great French mathematician
Henri Poincaré, had already formulated the essential
ideas and transformations several years before Einstein—although they had not appreciated the full
nature of the revolution, nor the virtue of taking the
relativity principle as axiomatic for physical forces
generally. (Apparently the British physicist Joseph Lar-
mor also had these transformations before both of
them, and there were partial results of this nature earlier still, by the German physicist Woldemar Voigt and
the Irishman George Francis FitzGerald.) Moreover, it
is my view that even Einstein had not fully understood
the nature of special relativity in 1905, as it was not
until 1908 that the final mathematical insights were
provided by the highly original Russian-German
geometer Hermann Minkowski (who had, coinciden-
tally, been one of Einstein's teachers at the Federal
Institute of Technology in Zurich in the late 1890s). It
was Minkowski's idea to combine time with space, and
to describe physical processes as inhabitants of the
four-dimensional space, now referred to as *space-time*.

It took Einstein some while to appreciate the virtue
of Minkowski's four-dimensional perspective. Even in
the account that he gives here, he betrays some unease
with Minkowski's viewpoint, emphasizing the analogy

with Euclidean geometry, but with an "imaginary" time-coordinate (i.e., measured in units involving $\sqrt{-1}$, which I regard as somewhat misleading) and he does not describe this space-time geometry in the way that I believe to be most physically direct. Physically realizable curves in space-time are *timelike*, and hence, as is reflected in more modern accounts, the metric is really supplying a measure of *time* to the curve, rather than length. In the excellent expositions by such relativity experts as John Leighton Synge and Hermann Bondi, it has been emphasized that space-time geometry is really *chronometry* rather than geometry in the ordinary sense, so that it is *clocks* that express the metric better than the little rulers that Einstein used in his descriptions. Rulers do *not* directly measure (spacelike) space-time intervals: because they require, in addition, some means of determination of "simultaneity," they are rather confusing in this regard. Clocks, however, *do* directly provide a measure of (timelike) space-time intervals, and they directly provide a good and superbly *accurate* measure the space-time geometry. (Indeed, the modern definition of a meter is now given in terms of a time measure, as precisely 1/299,792,458 of a light-second.) I would say that, in many respects, this book will be enjoyed as much for its historical insights into the time when the theory of relativity was young, as for its exposition of the very nature of Einstein's wonderful theories of relativity.

Einstein depended upon Minkowski's space-time idea for the generalizations that he needed in order to incorporate *gravity* into Minkowski's picture, whereby the *general* theory of relativity could be formulated. This theory was of a kind—almost unique in scientific development—where it is not unreasonable to suppose that, had Einstein not been there, the theory might well still not have been found by anyone, a century or more later. As I mentioned earlier, the work of many individuals contributed to special relativity, and the theory would still have come early in the twentieth century without Einstein. But the general theory required a different order of originality, and it is hard to see any plausible train of thought that would have clearly guided others to this theory.

Minkowski's geometry for special relativity was a "flat" one—basically a version of Euclid's ancient geometry of three-dimensional space, but one where Minkowski needed to modify Euclid's geometry, not just from three to four dimensions, but also in a way that is appropriate for a theory of space-time rather than just of space (using what we now call a "Lorentzian" metric).

Minkowski space provides the background for special relativity, but it does not incorporate gravity. What Einstein found that he needed was an *irregular* rather than a uniform geometry, so that Minkowski's geometry would become distorted in complicated ways, in order to describe all the details of a complicated gravitational field.

In fact, it was fortunate for Einstein that such irregular geometries had already been studied in pure mathematics (in the purely spatial context, rather than in the "Lorentzian" space-time one), in the nineteenth century, by the great German mathematicians Carl Friedrich Gauss and Georg Friedrich Bernhard Riemann. Gauss appears to have contemplated that physical space might be possibly be a non-flat (but uniform) geometry, and Riemann that the presence of matter might conceivably lead to a physical space with some kind of irregular geometry. Riemann's geometry formed an essential ingredient of Einstein's general relativity. But it was an insight of a quite different and very powerful kind that led Einstein to realize that the key ingredient to space-time curvature was *gravity*.

What was this insight? Einstein had already formulated special relativity, in which the physical laws were to be the same in any "inertial" frame of reference. In this theory, as in Newton's theory, an *inertial* frame is one in which the standard dynamical laws hold. As we move from one inertial frame to another, uniform motion in a straight line would translate to uniform motion in a straight line. The key to special relativity was that when we pass from one such frame to another, the speed of light must remain the same. Einstein was disturbed by this restriction to frames which are inertial, in this sense, and he tried to see how it could be possible for the physical laws to be the same, also, in a *non*-inertial (accelerating) frame. This issue was brilliantly solved by "the happiest thought" in Einstein's life, when he realized, while sitting in a chair in the Bern patent office, that "If a person falls freely he will not feel his own weight." In other words, the physics in an accelerating frame differs from that in an inertial one merely by an effective *gravitational field*. This he termed "the principle of equivalence," i.e., the equivalence of acceleration to a gravitational field. It had, in effect, been known to Galileo (all bodies fall with the same acceleration in a gravitational field), but had lain largely unappreciated for over three hundred years. The key idea of *general* relativity, where gravity now enters the picture, is that the notion of "inertial frame" needs to be reconsidered. Einstein's new notion of inertial frame differs from that of Newton. To Newton, a frame fixed on the Earth (ignoring the Earth's rotation) would be inertial; but to Einstein, such a frame would not be inertial. Instead, a *freely falling* frame, in the Earth's gravitational field, would count as inertial. Einstein's innovation, when gravitational fields are present, was to define the inertial ones to be where an observer feels no gravitational (or acceleration) force.

Einstein's ability to base a theory of gravity on this one key principle, together with another of his discoveries during the "Miraculous Year" of 1905, that mass and energy are basically the same thing ($E = mc^2$), and his realization that this would lead him to a theory based on curved space-time, was an astounding achievement—one matched only by the extraordinary precision that general relativity is now found to have (seen particularly in observations of the binary pulsar PSR1913+16), and by the utility that general-relativistic effects have in cosmology, through the effects of gravitational lensing. Even what Einstein considered to be his "greatest mistake," namely his introduction of a *cosmological constant* in 1917 (which had prevented him from predicting the expansion of the universe) has turned out to be a required ingredient of modern cosmology since 1998.

Einstein's considerations of cosmology, as described here, are fascinating. Prior to Einstein's theory, almost nothing reliable could be said about the structure of the universe as a whole. But Einstein was quick to realize that his general theory could indeed be applied to the entire universe. Although modern observations were not yet available, some of the mathematical constraints that the theory provided could now begin to be examined. We see here, indeed, the beginnings of the modern approach to cosmology.

Einstein's account is aimed at a level of technicality that makes it accessible to a general reader, and it is successful to a considerable degree. However, his descriptions in this book are minimal when it comes to the actual field equations of general relativity that he introduced in order to describe the gravitational field. This is in some contrast with his descriptions of special relativity, where Lorentz transformations are given explicitly. Whereas the metric quantities g_{ik} are referred to, the necessary (Riemann) space-time curvature tensor quantities are not included at all, being referred to only peripherally as providing the quantities whose vanishing would provide the conditions for entirely inertial reference frames to exist (flat space-time). Clearly Einstein felt that he could not describe such things in detail, in a book aimed at an entirely lay audience.

This is hardly surprising, particularly for a book written so soon after the discovery of general relativity. But the result was that the real nature of general relativity—which has to include the Einstein field equations—could not be fully conveyed. As I have said, he makes no attempt to explain to the reader the essential meaning of the notion of "space-time curvature," which is the mathematical key to his theory. Modern popular descriptions can take full advantage of the physical manifestation of space-time curvature in terms of tidal distortions (as pointed out in the 1950s by Felix Pirani and others), but these were not available to Einstein at the time, and his account inevitably falls short of providing any such physical interpretation of space-time curvature. In 1922, he published *The Meaning of Relativity*, which, though still written at a popular level, contains more technical detail than the present work.

Relativity is perhaps Einstein's most widely known book. Although he clearly demonstrates a bold confidence when describing his actual physical ideas, I feel that he exhibits a certain surprising insecurity in his actual ability to convey, in an elegant way, these ideas in words. At the beginning of the book, he quotes with approval a comment made by the great physicist Ludwig Boltzmann that "matters of elegance ought to be left to the tailor and to the cobbler." Yet there is a definite elegance in Einstein's writings; when at its fullest, such elegance makes the reading easier. Nevertheless, there is a certain unevenness in the writing; the passages which flow most smoothly are those where Einstein is most comfortable with the concepts he is describing. He is perhaps at his most awkward (and even repetitive) when describing the roles of coordinate systems and frames of reference, using the curious and confusing analogy of a mollusc (or "reference mollusc"—his German word being "Bezugsmolluske"), indicating that he apparently thought of a coordinate system as a physical thing, swimming around like some soft marine invertebrate which could change its shape with time. (It took the mathematician Hermann Weyl to provide the modern viewpoint with regard to coordinates; see the fifth edition of his *Space-Time-Matter* [1923].) But he is at his most inspiring when describing his powerful physical ideas and his novel global insights, like the principle of equivalence and his speculations concerning the global nature of the universe.

The present edition includes material that was added in the 1920 English translation, but not the more recent material of the enlarged edition published in January 1954, the year before Einstein died, in which he made a brief mention of his later ideas for obtaining a "unified field theory," according to which electromagnetism as well as gravity were to be described. It has to be said that these grand ideas that Einstein played with at the end of his life did not really amount to anything of unambiguous significance. It is regarded as a particular weakness of Einstein's approaches to the formulation of a unified field theory that he appeared to make no attempt to incorporate physical interactions other than gravitational or elec-tromagnetic into his unified schemes. Einstein is also commonly regarded as having lost touch, in his later years, with the developments of modern physics—a quantum physics that he had himself been instrumen-tal in initiating—through his inability to accept the emergent tenets of the resulting quantum-mechanical theory. Yet, in my opinion, there is some deep truth in Einstein's difficulties with the subjectivity and lack of realism in the way that quantum theory has developed.

I would concur with him that modern quantum theory is an incomplete theory, and that new developments must be waiting in the wings that may well supply us with a new quantum revolution, perhaps comparable with Einstein's own general relativity. Indeed, it is my personal view that the very insights that drew Einstein to formulate general relativity, most particularly the principle of equivalence, may supply one of the vital links to such a theory. But this is a matter for the future, and perhaps a *new* Einstein!

Since Einstein's death in 1955, there have been many developments of importance in the subject of general relativity. Perhaps the most striking of these have been on the observational side, which have led to impressively accurate confirmations of the profound accord that Einstein's general theory finds with the workings of Nature. This started in 1960, when R. V. Pound and G. A. Rebka were able to provide a convincing confirmation of Einstein's prediction that time is slightly slowed down in a gravitational field. This was one of Einstein's famous "three tests" of general relativity—which was, unfortunately, *not* really "definitely established by Adams in 1924, by observations on the dense companion of Sirius," as optimistically remarked in a footnote by Einstein's translator at the end of Appendix 3 to *Relativity*. Pound and Rebka achieved this by comparing time rates at the bottom and the top of a 22.5 meter tower, where the rate at the top was greater by a proportion that would amount to a mere second in 30 million years!

In Appendix 3, Einstein reports on the results Arthur Eddington's now famous expedition to the island of Principe in 1919, which provided a rough confirmation of the general-relativistic prediction that the sun's gravitational field would deflect light (another of the "three tests"). This has now been vastly improved upon, starting with quasar radio observations in 1969. Such "gravitational lensing" effects are now used as an important observational tool in cosmology, giving a direct determination of very distant distributions of mass.

The most impressive confirmation of Einstein's theory during his lifetime (the remaining one of the "three tests") was an apparently anomalous orbital precession of the orbit of the planet Mercury. Such effects, together with a new "time delay effect" are now confirmed to great precision in the solar system as a whole in work by Irwin Shapiro and his colleagues. To date, the most impressive test of all is in the orbital motions in the double neutron-star system PSR

1913+16, whose precision over a period of about thirty years of observation was timed, by the Nobel-prize winning team of Joseph Taylor and Russell Hulse, to an accuracy of about one part in a hundred million million, where many general-relativistic effects contribute to this in essential ways, including the energy loss from the system due to its emission of gravitational waves. It may also be mentioned that, at a more familiar level of application, the effects of general relativity must be taken into account for the successful operation of GPS devices.

There has been much progress also on the theoretical side. Perhaps the most impressive of these it the theory of black holes which, despite the early work of Chandrasekhar (1931), Oppenheimer and Snyder (1939), and others, during Einstein's lifetime, seems not to have been taken really seriously by him. We now know that gravitational collapse to a black hole can occur without any assumption of symmetry having to be made, and where there is no particular restriction on the type of material composition. All that is needed is sufficient material concentrated within a small enough region. This can happen to individual stars (of, say, ten times the mass of the sun) or in large collections of stars at galactic centers, and now there is good observational evidence for black holes arising in both types of situation. Theoretical considerations tell us that at the central region of a black hole, Einstein's classical field equations (where "classical" means in the absence of quantum mechanical considerations) must reach their limit, leading to what we refer to as a "space-time singularity," where matter densities and space-time curvatures are expected to attain infinite values. This troublesome region is, however, anticipated to be surrounded by an "event horizon," which would prevent particles or signals of any kind from reaching the outside. Yet, in the general case, the necessary existence of such a "horizon" remains an unproved conjecture—referred to as "cosmic censorship"—arguably the most important unsolved mathematical problem of classical general relativity. On the other hand, it has been established that provided such an event horizon indeed occurs, and if it is assumed that the space-time is stationary, then we obtain a very specific black-hole geometry which had been obtained explicitly by Roy Kerr in 1963. As the distinguished astrophysicist Subramanyan Chandrasekhar has remarked, black holes are the most perfect macroscopic objects in the universe, with a mathematical description which is unexpectedly simple and elegant.

There is an important physical quantity called *entropy*, possessed by physical systems, which may be regarded as a measure of how "probable" that system is to come about by chance. The fundamental *second law of thermodynamics* asserts that the entropy of an isolated system *increases* with time (or perhaps remains constant). A black hole has the remarkable property, as shown by Jakob Bekenstein and Stephen Hawking, that its entropy is a specific multiple of the surface area of its horizon. This entropy turns out to be enormous and, because of the large black holes at the centers of galaxies, this black-hole entropy easily represents the major contribution to the entropy of the known universe. Because of the second law, the entropy at the beginning of the universe must, on the other hand, be extremely small. This beginning is represented, according to the field equations of Einstein's general relativity, by another space-time singularity known as the *Big Bang*, which must, therefore, be extraordinarily special, or "fine tuned," as compared with the high-entropy singularities that lie at the cores of black holes. A satisfactory theory of the physics that takes place at those regions described, classically, as "space-time singularities" would have to go beyond Einstein's classical general relativity, and such a theory would normally be referred to as *quantum gravity*. It would appear to be a prime task of such a quantum-gravity theory to explain the remarkable time asymmetry between the singularities in black holes and that at the Big Bang, but despite half a century of concerted work, there is yet no broadly accepted theory of quantum gravity.

Finally, there is the remarkable observational conclusion of 1998, now largely accepted by the community of cosmologists, that the *cosmological constant* that Einstein introduced in 1917, but which he subsequently withdrew as his "greatest mistake," actually has a small but significant (positive) value. This has the implication that the remote future of the universe will be an (unexpected) exponential expansion. This remains something of a puzzle to theoreticians, but it represents one of the many challenges to be faced by cosmologists in the future, and exemplifies the extraordinary richness that has been opened up by Einstein's fantastic world-view, according to general relativity, of a universe governed, up to its greatest scales, by a four-dimensional curved space-time geometry.

NOTE ON THE TEXT

Einstein's preface, thirty-two chapters, and three appendixes that make up the text of the Masterpiece Science edition of *Relativity* were reproduced from the original 1920 edition. Spelling errors were corrected, and all figures were redrawn.

A † following a chapter title indicates that new commentary for that chapter can be found in Robert Geroch's commentary section, which begins after the third appendix of Einstein's text.

RELATIVITY

The Special and General Theory

PREFACE†

The present book is intended, as far as possible, to give an exact insight into the theory of Relativity to those readers who, from a general scientific and philosophical point of view, are interested in the theory, but who are not conversant with the mathematical apparatus of theoretical physics. The work presumes a standard of education corresponding to that of a university matriculation examination, and, despite the shortness of the book, a fair amount of patience and force of will on the part of the reader. The author has spared himself no pains in his endeavour to present the main ideas in the simplest and most intelligible form, and on the whole, in the sequence and connection in which they actually originated. In the interest of clearness, it appeared to me inevitable that I should repeat myself frequently, without paying the slightest attention to the elegance of the presentation. I adhered scrupulously to the precept of that brilliant theoretical physicist, L. Boltzmann, according to whom matters of elegance ought to be left to the tailor and to the cobbler. I make no pretence of having with-held from the reader difficulties which are inherent to the subject. On the other hand, I have purposely treated the empirical physical foundations of the theory in a "step-motherly" fashion, so that readers unfamiliar with physics may not feel like the wanderer who was unable to see the forest for trees. May the book bring some one a few happy hours of suggestive thought!

A. EINSTEIN

December, 1916

PART I

THE SPECIAL THEORY
OF RELATIVITY

1

PHYSICAL MEANING OF
GEOMETRICAL PROPOSITIONS

In your schooldays most of you who read this book made acquaintance with the noble building of Euclid's geometry, and you remember—perhaps with more respect than love—the magnificent structure, on the lofty staircase of which you were chased about for uncounted hours by conscientious teachers. By reason of your past experience, you would certainly regard every one with disdain who should pronounce even the most out-of-the-way proposition of this science to be untrue. But perhaps this feeling of proud certainty would leave you immediately if some one were to ask you: "What, then, do you mean by the assertion that these propositions are true?" Let us proceed to give this question a little consideration.

Geometry sets out from certain conceptions such as "plane," "point," and "straight line," with which we are able to associate more or less definite ideas, and from certain simple propositions (axioms) which, in virtue of these ideas, we are inclined to accept as "true." Then, on the basis of a logical process, the justification of which we feel ourselves compelled to admit, all remaining propositions are shown to follow from those axioms, *i.e.* they are proven. A proposition is then correct ("true") when it has been derived in the recognised manner from the axioms. The question of the "truth" of the individual geometrical propositions is thus reduced to one of the "truth" of the axioms. Now it has long been known that the last question is not only unanswerable by the methods of geometry, but that it is in itself entirely without meaning. We cannot ask whether it is true that only one straight line goes through two points. We can only say that Euclidean geometry deals with things called "straight lines," to each of which is ascribed the property of being uniquely determined by two points situated on it. The concept "true" does not tally with the assertions of pure geometry, because by the word "true" we are eventually in the habit of designating always the correspondence with a "real" object; geometry, however, is not concerned with the relation of the ideas involved in it to objects of experience, but only with the logical connection of these ideas among themselves.

It is not difficult to understand why, in spite of this, we feel constrained to call the propositions of geometry "true." Geometrical ideas correspond to more or less exact objects in nature, and these last are undoubtedly the exclusive cause of the genesis of those ideas. Geometry ought to refrain from such a course, in order to give to its structure the largest possible logical unity. The practice, for example, of seeing in a "distance" two marked positions on a practically rigid body is something which is lodged deeply in our habit of thought. We are accustomed further to regard three points as being situated on a straight line, if their apparent positions can be made to coincide for observation with one eye, under suitable choice of our place of observation.

If, in pursuance of our habit of thought, we now supplement the propositions of Euclidean geometry by the single proposition that two points on a practically rigid body always correspond to the same distance (line-interval), independently of any changes in position to which we may subject the body, the propositions of Euclidean geometry then resolve themselves into propositions on the possible relative position of practically rigid bodies.[1] Geometry which has been supplemented in this way is then to be treated as a branch of physics. We can now legitimately ask as to the "truth" of geometrical propositions interpreted in this way, since we are justified in asking whether these propositions are satisfied for those real things we have associated with the geometrical ideas. In less exact terms we can express this by saying that by the "truth" of a geometrical proposition in this sense we understand its validity for a construction with ruler and compasses.

Of course the conviction of the "truth" of geometrical propositions in this sense is founded exclusively on rather incomplete experience. For the present we shall assume the "truth" of the geometrical propositions, then at a later stage (in the general theory of relativity) we shall see that this "truth" is limited, and we shall consider the extent of its limitation.

---

1. It follows that a natural object is associated also with a straight line. Three points A, B and C on a rigid body thus lie in a straight line when, the points A and C being given, B is chosen such that the sum of the distances AB and BC is as short as possible. This incomplete suggestion will suffice for our present purpose.

2

THE SYSTEM OF CO-ORDINATES

O n the basis of the physical interpretation of distance which has been indicated, we are also in a position to establish the distance between two points on a rigid body by means of measurements. For this purpose we require a "distance" (rod S) which is to be used once and for all, and which we employ as a standard measure. If, now, A and B are two points on a rigid body, we can construct the line joining them according to the rules of geometry; then, starting from A, we can mark off the distance S time after time until we reach B. The number of these operations required is the numerical measure of the distance AB. This is the basis of all measurement of length.[1]

Every description of the scene of an event or of the position of an object in space is based on the specification of the point on a rigid body (body of reference) with which that event or object coincides. This applies not only to scientific description, but also to everyday

1. Here we have assumed that there is nothing left over, *i.e.* that the measurement gives a whole number. This difficulty is got over by the use of divided measuring-rods, the introduction of which does not demand any fundamentally new method.

life. If I analyse the place specification "Trafalgar Square, London," [2] I arrive at the following result. The earth is the rigid body to which the specification of place refers; "Trafalgar Square, London"[1] is a well-defined point, to which a name has been assigned, and with which the event coincides in space.[3]

This primitive method of place specification deals only with places on the surface of rigid bodies, and is dependent on the existence of points on this surface which are distinguishable from each other. But we can free ourselves from both of these limitations without altering the nature of our specification of position. If, for instance, a cloud is hovering over Trafalgar Square, then we can determine its position relative to the surface of the earth by erecting a pole perpendicularly on the Square, so that it reaches the cloud. The length of the pole measured with the standard measuring-rod, combined with the specification of the position of the foot of the pole, supplies us with a complete place specification. On the basis of this illustration, we are able to see the manner in which a refinement of the conception of position has been developed.

2. I have chosen this as being more familiar to the English reader than the "Potsdamer Platz, Berlin," which is referred to in the original. (R. W. L.)

3. It is not necessary here to investigate further the significance of the expression "coincidence in space." This conception is sufficiently obvious to ensure that differences of opinion are scarcely likely to arise as to its applicability in practice.

(*a*) We imagine the rigid body, to which the place specification is referred, supplemented in such a manner that the object whose position we require is reached by the completed rigid body.

(*b*) In locating the position of the object, we make use of a number (here the length of the pole measured with the measuring-rod) instead of designated points of reference.

(*c*) We speak of the height of the cloud even when the pole which reaches the cloud has not been erected. By means of optical observations of the cloud from different positions on the ground, and taking into account the properties of the propagation of light, we determine the length of the pole we should have required in order to reach the cloud.

From this consideration we see that it will be advantageous if, in the description of position, it should be possible by means of numerical measures to make ourselves independent of the existence of marked positions (possessing names) on the rigid body of reference. In the physics of measurement this is attained by the application of the Cartesian system of co-ordinates.

This consists of three plane surfaces perpendicular to each other and rigidly attached to a rigid body. Referred to a system of co-ordinates, the scene of any event will be determined (for the main part) by the specification of the lengths of the three perpendiculars or co-ordinates (x, y, z) which can be dropped from the scene of the event to those three plane surfaces. The lengths of these three perpendiculars can be determined by a series of manipulations with rigid measuring-rods performed according to the rules and methods laid down by Euclidean geometry.

In practice, the rigid surfaces which constitute the system of co-ordinates are generally not available; furthermore, the magnitudes of the co-ordinates are not actually determined by constructions with rigid rods, but by indirect means. If the results of physics and astronomy are to maintain their clearness, the physical meaning of specifications of position must always be sought in accordance with the above considerations.[4]

We thus obtain the following result: Every description of events in space involves the use of a rigid body to which such events have to be referred. The resulting relationship takes for granted that the laws of Euclidean geometry hold for "distances," the "distance" being represented physically by means of the convention of two marks on a rigid body.

---

4. A refinement and modification of these views does not become necessary until we come to deal with the general theory of relativity, treated in the second part of this book.

3

SPACE AND TIME IN CLASSICAL MECHANICS

"The purpose of mechanics is to describe how bodies change their position in space with time." I should load my conscience with grave sins against the sacred spirit of lucidity were I to formulate the aims of mechanics in this way, without serious reflection and detailed explanations. Let us proceed to disclose these sins.

It is not clear what is to be understood here by "position" and "space." I stand at the window of a railway carriage which is travelling uniformly, and drop a stone on the embankment, without throwing it. Then, disregarding the influence of the air resistance, I see the stone descend in a straight line. A pedestrian who observes the misdeed from the footpath notices that the stone falls to earth in a parabolic curve. I now ask: Do the "positions" traversed by the stone lie "in reality" on a straight line or on a parabola? Moreover, what is meant here by motion "in space"? From the considerations of the previous section the answer is self-evident. In the first place, we entirely shun the vague word "space," of which, we must honestly acknowledge, we cannot form the slightest conception, and we replace it by "motion relative to a practically rigid body of reference." The positions relative to the body of reference (railway carriage or embankment) have already been defined in detail in the preceding section. If instead of "body of reference" we insert "system of co-ordinates," which is a useful idea for mathematical description, we are in a position to say: The stone traverses a straight line relative to a system of co-ordinates rigidly attached to the carriage, but relative to a system of co-ordinates rigidly attached to the ground (embankment) it describes a parabola. With the aid of this example it is clearly seen that there is no such thing as an independently existing trajectory (lit. "path-curve"[1]), but only a trajectory relative to a particular body of reference.

In order to have a *complete* description of the motion, we must specify how the body alters its position *with time; i.e.* for every point on the trajectory it must be stated at what time the body is situated there. These data must be supplemented by such a definition of time that, in virtue of this definition, these time-values can be regarded essentially as magnitudes (results of measure-

---

1. That is, a curve along which the body moves.

ments) capable of observation. If we take our stand on the ground of classical mechanics, we can satisfy this requirement for our illustration in the following manner. We imagine two clocks of identical construction; the man at the railway-carriage window is holding one of them, and the man on the footpath the other. Each of the observers determines the position on his own reference-body occupied by the stone at each tick of the clock he is holding in his hand. In this connection we have not taken account of the inaccuracy involved by the finiteness of the velocity of propagation of light. With this and with a second difficulty prevailing here we shall have to deal in detail later.

4

THE GALILEIAN SYSTEM OF CO-ORDINATES

A s is well known, the fundamental law of the mechanics of Galilei-Newton, which is known as the *law of inertia,* can be stated thus: A body removed sufficiently far from other bodies continues in a state of rest or of uniform motion in a straight line. This law not only says something about the motion of the bodies, but it also indicates the reference-bodies or systems of co-ordinates, permissible in mechanics, which can be used in mechanical description. The visible fixed stars are bodies for which the law of inertia certainly holds to a high degree of approximation. Now if we use a system of co-ordinates which is rigidly attached to the earth, then, relative to this system, every fixed star describes a circle of immense radius in the course of an astronomical day, a result which is opposed to the statement of the law of inertia. So that if we adhere to this law we must refer these motions only to systems of co-ordinates relative to which the fixed stars do not move in a circle. A system of co-ordinates of which the state of motion is such that the law of inertia holds relative to it is called a "Galileian system of co-ordinates." The laws of the mechanics of Galilei-Newton can be regarded as valid only for a Galileian system of co-ordinates.

5

THE PRINCIPLE OF RELATIVITY
(IN THE RESTRICTED SENSE)†

In order to attain the greatest possible clearness, let us return to our example of the railway carriage supposed to be travelling uniformly. We call its motion a uniform translation ("uniform" because it is of constant velocity and direction, "translation" because although the carriage changes its position relative to the embankment yet it does not rotate in so doing). Let us imagine a raven flying through the air in such a manner that its motion, as observed from the embankment, is uniform and in a straight line. If we were to observe the flying raven from the moving railway carriage, we should find that the motion of the raven would be one of different velocity and direction, but that it would still be uniform and in a straight line. Expressed in an abstract manner we may say: If a mass m is moving uniformly in a straight line with respect to a co-ordinate system K, then it will also be moving uniformly and in a straight line relative to a second co-ordinate system K', provided that the latter is executing a uniform translatory motion with respect to K.

In accordance with the discussion contained in the preceding section, it follows that:

If K is a Galileian co-ordinate system, then every other co-ordinate system K' is a Galileian one, when, in relation to K, it is in a condition of uniform motion of translation. Relative to K' the mechanical laws of Galilei-Newton hold good exactly as they do with respect to K.

We advance a step farther in our generalisation when we express the tenet thus: If, relative to K, K' is a uniformly moving co-ordinate system devoid of rotation, then natural phenomena run their course with respect to K' according to exactly the same general laws as with respect to K. This statement is called the *principle of relativity* (in the restricted sense).

As long as one was convinced that all natural phenomena were capable of representation with the help of classical mechanics, there was no need to doubt the validity of this principle of relativity. But in view of the more recent development of electrodynamics and optics it became more and more evident that classical mechanics affords an insufficient foundation for the physical description of all natural phenomena. At this juncture the question of the validity of the principle of relativity became ripe for discussion, and it did not appear impossible that the answer to this question might be in the negative.

Nevertheless, there are two general facts which at the outset speak very much in favour of the validity of the principle of relativity. Even though classical mechanics does not supply us with a sufficiently broad basis for the theoretical presentation of all physical phenomena, still we must grant it a considerable measure of "truth," since it supplies us with the actual motions of the heavenly bodies with a delicacy of detail little short of wonderful. The principle of relativity must therefore apply with great accuracy in the domain of *mechanics*. But that a principle of such broad generality should hold with such exactness in one domain of phenomena, and yet should be invalid for another, is *a priori* not very probable.

We now proceed to the second argument, to which, moreover, we shall return later. If the principle of relativity (in the restricted sense) does not hold, then the Galileian co-ordinate systems K, K', K'', etc., which are moving uniformly relative to each other, will not be *equivalent* for the description of natural phenomena. In this case we should be constrained to believe that natural laws are capable of being formulated in a particularly simple manner, and of course only on condition that, from amongst all possible Galileian co-ordinate systems, we should have chosen *one* (K_0) of a particular state of motion as our body of reference. We should then be justified (because of its merits for the description of natural phenomena) in calling this system "absolutely at rest," and all other Galileian systems K "in motion." If, for instance, our embankment were the system K_0, then our railway carriage would be a system K, relative to which less simple laws would hold than with respect to K_0. This diminished simplicity would be due to the fact that the carriage K would be in motion (*i.e.* "really") with respect to K_0. In the general laws of nature which have been formulated with reference to K, the magnitude and direction of the velocity of the carriage would necessarily play a part. We should expect, for instance, that the note emitted by an organ-pipe placed with its axis parallel to the direction of travel would be different from that emitted if the axis of the pipe were placed perpendicular to this direction. Now in virtue of its motion in an orbit round the sun, our earth is comparable with a railway carriage travelling with a velocity of about 30 kilometres per second. If the principle of relativity were not valid we should therefore expect that the direction of motion of the earth at any moment would enter into the laws of nature, and also that physical systems in their behaviour would be dependent on the orientation in space with respect to the earth. For owing to the alteration in direction of the velocity of rotation of the earth in the course of a year, the earth cannot be at rest relative to the hypothetical system K_0 throughout the whole year. However, the most careful observations have never revealed such anisotropic properties in terrestrial physical space, *i.e.* a physical non-equivalence of different directions. This is a very powerful argument in favour of the principle of relativity.

6

THE THEOREM OF THE ADDITION OF VELOCITIES EMPLOYED IN CLASSICAL MECHANICS

Let us suppose our old friend the railway carriage to be travelling along the rails with a constant velocity v, and that a man traverses the length of the carriage in the direction of travel with a velocity w. How quickly, or, in other words, with what velocity W does the man advance relative to the embankment during the process? The only possible answer seems to result from the following consideration: If the man were to stand still for a second, he would advance relative to the embankment through a distance v equal numerically to the velocity of the carriage. As a consequence of his walking, however, he traverses an additional distance w relative to the carriage, and hence also relative to the embankment, in this second, the distance w being numerically equal to the velocity with which he is walking. Thus in total he covers the distance $W = v + w$ relative to the embankment in the second considered.

We shall see later that this result, which expresses the theorem of the addition of velocities employed in classical mechanics, cannot be maintained; in other words, the law that we have just written down does not hold in reality. For the time being, however, we shall assume its correctness.

7

THE APPARENT INCOMPATIBILITY OF THE LAW OF PROPAGATION OF LIGHT WITH THE PRINCIPLE OF RELATIVITY†

There is hardly a simpler law in physics than that according to which light is propagated in empty space. Every child at school knows, or believes he knows, that this propagation takes place in straight lines with a velocity $c = 300,000$ km./sec. At all events we know with great exactness that this velocity is the same for all colours, because if this were not the case, the minimum of emission would not be observed simultaneously for different colours during the eclipse of a fixed star by its dark neighbour. By means of similar considerations based on observations of double stars, the Dutch astronomer De Sitter was also able to show that the velocity of propagation of light cannot depend on the velocity of motion of the body emitting the light. The assumption that this velocity of propagation is dependent on the direction "in space" is in itself improbable.

In short, let us assume that the simple law of the constancy of the velocity of light c (in vacuum) is justifiably believed by the child at school. Who would imagine that this simple law has plunged the conscientiously thoughtful physicist into the greatest intellectual difficulties? Let us consider how these difficulties arise.

Of course we must refer the process of the propagation of light (and indeed every other process) to a rigid reference-body (co-ordinate system). As such a system let us again choose our embankment. We shall imagine the air above it to have been removed. If a ray of light be sent along the embankment, we see from the above that the tip of the ray will be transmitted with the velocity c relative to the embankment. Now let us suppose that our railway carriage is again travelling along the railway lines with the velocity v, and that its direction is the same as that of the ray of light, but its velocity of course much less. Let us inquire about the velocity of propagation of the ray of light relative to the carriage. It is obvious that we can here apply the consideration of the previous section, since the ray of light plays the part of the man walking along relatively to the carriage. The velocity W of the man relative to the embankment is here replaced by the velocity of light relative to the embankment. w is the required velocity of light with respect to the carriage, and we have

$$w = c - v.$$

The velocity of propagation of a ray of light relative to the carriage thus comes out smaller than *c*.

But this result comes into conflict with the principle of relativity set forth in Section 5. For, like every other general law of nature, the law of the transmission of light *in vacuo* must, according to the principle of relativity, be the same for the railway carriage as reference-body as when the rails are the body of reference. But, from our above consideration, this would appear to be impossible. If every ray of light is propagated relative to the embankment with the velocity *c*, then for this reason it would appear that another law of propagation of light must necessarily hold with respect to the carriage—a result contradictory to the principle of relativity.

In view of this dilemma there appears to be nothing else for it than to abandon either the principle of relativity or the simple law of the propagation of light *in vacuo*. Those of you who have carefully followed the preceding discussion are almost sure to expect that we should retain the principle of relativity, which appeals so convincingly to the intellect because it is so natural and simple. The law of the propagation of light *in vacuo* would then have to be replaced by a more complicated law conformable to the principle of relativity. The development of theoretical physics shows, however, that we cannot pursue this course. The epoch-making theoretical investigations of H. A. Lorentz on the electrodynamical and optical phenomena connected with moving bodies show that experience in this domain leads conclusively to a theory of electromagnetic phenomena, of which the law of the constancy of the velocity of light *in vacuo* is a necessary consequence. Prominent theoretical physicists were therefore more inclined to reject the principle of relativity, in spite of the fact that no empirical data had been found which were contradictory to this principle.

At this juncture the theory of relativity entered the arena. As a result of an analysis of the physical conceptions of time and space, it became evident that *in reality there is not the least incompatibility between the principle of relativity and the law of propagation of light,* and that by systematically holding fast to both these laws a logically rigid theory could be arrived at. This theory has been called the *special theory of relativity* to distinguish it from the extended theory, with which we shall deal later. In the following pages we shall present the fundamental ideas of the special theory of relativity.

8

ON THE IDEA OF TIME
IN PHYSICS

Lightning has struck the rails on our railway embankment at two places *A* and *B* far distant from each other. I make the additional assertion that these two lightning flashes occurred simultaneously. If now I ask you whether there is sense in this statement, you will answer my question with a decided "Yes." But if I now approach you with the request to explain to me the sense of the statement more precisely, you find after some consideration that the answer to this question is not so easy as it appears at first sight.

After some time perhaps the following answer would occur to you: "The significance of the statement is clear in itself and needs no further explanation; of course it would require some consideration if I were to be commissioned to determine by observations whether in the actual case the two events took place simultaneously or not." I cannot be satisfied with this answer for the following reason. Supposing that as a result of ingenious considerations an able meteorologist were to discover that the lightning must always strike the places *A* and *B* simultaneously, then we should be faced with the task of testing whether or not this theoretical result is in accordance with the reality. We encounter the same difficulty with all physical statements in which the conception "simultaneous" plays a part. The concept does not exist for the physicist until he has the possibility of discovering whether or not it is fulfilled in an actual case. We thus require a definition of simultaneity such that this definition supplies us with the method by means of which, in the present case, he can decide by experiment whether or not both the lightning strokes occurred simultaneously. As long as this requirement is not satisfied, I allow myself to be deceived as a physicist (and of course the same applies if I am not a physicist), when I imagine that I am able to attach a meaning to the statement of simultaneity. (I would ask the reader not to proceed farther until he is fully convinced on this point.)

After thinking the matter over for some time you then offer the following suggestion with which to test simultaneity. By measuring along the rails, the connecting line *AB* should be measured up and an observer placed at the mid-point *M* of the distance *AB*. This observer should be supplied with an arrangement (*e.g.* two mirrors inclined at 90°) which allows him visually to observe both places A and B at the same time. If the observer perceives the two flashes of lightning at the same time, then they are simultaneous.

I am very pleased with this suggestion, but for all that I cannot regard the matter as quite settled, because I feel constrained to raise the following objection: "Your definition would certainly be right, if I only knew that the light by means of which the observer at M perceives the lightning flashes travels along the length $A \longrightarrow M$ with the same velocity as along the length $B \longrightarrow M$. But an examination of this supposition would only be possible if we already had at our disposal the means of measuring time. It would thus appear as though we were moving here in a logical circle."

After further consideration you cast a somewhat disdainful glance at me—and rightly so—and you declare: "I maintain my previous definition nevertheless, because in reality it assumes absolutely nothing about light. There is only *one* demand to be made of the definition of simultaneity, namely, that in every real case it must supply us with an empirical decision as to whether or not the conception that has to be defined is fulfilled. That my definition satisfies this demand is indisputable. That light requires the same time to traverse the path $A \longrightarrow M$ as for the path $B \longrightarrow M$ is in reality neither a *supposition nor a hypothesis* about the physical nature of light, but a *stipulation* which I can make of my own freewill in order to arrive at a definition of simultaneity."

It is clear that this definition can be used to give an exact meaning not only to *two* events, but to as many events as we care to choose, and independently of the positions of the scenes of the events with respect to the body of reference[1] (here the railway embankment). We are thus led also to a definition of "time" in physics. For this purpose we suppose that clocks of identical construction are placed at the points *A*, *B* and *C* of the railway line (co-ordinate system), and that they are set in such a manner that the positions of their pointers are simultaneously (in the above sense) the same. Under these conditions we understand by the "time" of an event the reading (position of the hands) of that one of these clocks which is in the immediate vicinity (in space) of the event. In this manner a time-value is associated with every event which is essentially capable of observation.

---

1. We suppose further that, when three events *A*, *B* and *C* take place in different places in such a manner that, if *A* is simultaneous with *B*, and *B* is simultaneous with *C* (simultaneous in the sense of the above definition), then the criterion for the simultaneity of the pair of events *A*, *C* is also satisfied. This assumption is a physical hypothesis about the law of propagation of light; it must certainly be fulfilled if we are to maintain the law of the constancy of the velocity of light *in vacuo*.

This stipulation contains a further physical hypothesis, the validity of which will hardly be doubted without empirical evidence to the contrary. It has been assumed that all these clocks go *at the same rate* if they are of identical construction. Stated more exactly: When two clocks arranged at rest in different places of a reference-body are set in such a manner that a *particular* position of the pointers of the one clock is *simultaneous* (in the above sense) with the *same* position of the pointers of the other clock, then identical "settings" are always simultaneous (in the sense of the above definition).

9

THE RELATIVITY OF SIMULTANEITY

Up to now our considerations have been referred to a particular body of reference, which we have styled a "railway embankment." We suppose a very long train travelling along the rails with the constant velocity v and in the direction indicated in Fig. 1. People travelling in this train will with advantage use the train as a rigid reference-body (co-ordinate system); they regard all events in reference to

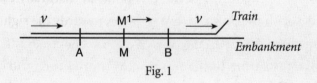

Fig. 1

the train. Then every event which takes place along the line also takes place at a particular point of the train. Also the definition of simultaneity can be given relative to the train in exactly the same way as with respect to the embankment. As a natural consequence, however, the following question arises:

Are two events (*e.g.* the two strokes of lightning *A* and *B*) which are simultaneous *with reference to the railway embankment* also simultaneous *relatively to the train?* We shall show directly that the answer must be in the negative.

When we say that the lightning strokes *A* and *B* are simultaneous with respect to the embankment, we mean: the rays of light emitted at the places *A* and *B*, where the lightning occurs, meet each other at the midpoint *M* of the length *A* —> *B* of the embankment. But the events *A* and *B* also correspond to positions *A* and *B* on the train. Let *M'* be the mid-point of the distance *A* —> *B* on the travelling train. Just when the flashes[1] of lightning occur, this point *M'* naturally coincides with the point *M*, but it moves towards the right in the diagram with the velocity *v* of the train. If an observer sitting in the position *M'* in the train did not possess this velocity, then he would remain permanently at *M*, and the light rays emitted by the flashes of lightning *A* and *B* would reach him simultaneously, *i.e.* they would meet just where he is situated. Now in reality (considered with reference to the railway embankment) he is hastening towards the beam of light coming from

1. As judged from the embankment.

B, whilst he is riding on ahead of the beam of light coming from *A.* Hence the observer will see the beam of light emitted from *B* earlier than he will see that emitted from *A.* Observers who take the railway train as their reference-body must therefore come to the conclusion that the lightning flash *B* took place earlier than the lightning flash *A.* We thus arrive at the important result:

Events which are simultaneous with reference to the embankment are not simultaneous with respect to the train, and *vice versa* (relativity of simultaneity). Every reference-body (co-ordinate system) has its own particular time; unless we are told the reference-body to which the statement of time refers, there is no meaning in a statement of the time of an event.

Now before the advent of the theory of relativity it had always tacitly been assumed in physics that the statement of time had an absolute significance, *i.e.* that it is independent of the state of motion of the body of reference. But we have just seen that this assumption is incompatible with the most natural definition of simultaneity; if we discard this assumption, then the conflict between the law of the propagation of light *in vacuo* and the principle of relativity (developed in Section 7) disappears.

We were led to that conflict by the considerations of Section 6, which are now no longer tenable. In that section we concluded that the man in the carriage, who traverses the distance *w per second* relative to the carriage, traverses the same distance also with respect to the embankment *in each second* of time. But, according to the foregoing considerations, the time required by a particular occurrence with respect to the carriage must not be considered equal to the duration of the same occurrence as judged from the embankment (as reference-body). Hence it cannot be contended that the man in walking travels the distance *w* relative to the railway line in a time which is equal to one second as judged from the embankment.

Moreover, the considerations of Section 6 are based on yet a second assumption, which, in the light of a strict consideration, appears to be arbitrary, although it was always tacitly made even before the introduction of the theory of relativity.

ON THE RELATIVITY OF THE CONCEPTION OF DISTANCE†

Let us consider two particular points on the train[1] travelling along the embankment with the velocity v, and inquire as to their distance apart. We already know that it is necessary to have a body of reference for the measurement of a distance, with respect to which body the distance can be measured up. It is the simplest plan to use the train itself as the reference-body (co-ordinate system). An observer in the train measures the interval by marking off his measuring-rod in a straight line (*e.g.* along the floor of the carriage) as many times as is necessary to take him from the one marked point to the other. Then the number which tells us how often the rod has to be laid down is the required distance.

It is a different matter when the distance has to be judged from the railway line. Here the following method suggests itself. If we call A' and B' the two points on the train whose distance apart is required, then both of these points are moving with the velocity v along the embankment.

1. *e.g.* the middle of the first and of the hundredth carriage.

In the first place we require to determine the points A and B of the embankment which are just being passed by the two points A' and B' at a particular time t—judged from the embankment. These points A and B of the embankment can be determined by applying the definition of time given in Section 8. The distance between these points A and B is then measured by repeated application of the measuring-rod along the embankment.

A priori it is by no means certain that this last measurement will supply us with the same result as the first. Thus the length of the train as measured from the embankment may be different from that obtained by measuring in the train itself. This circumstance leads us to a second objection which must be raised against the apparently obvious consideration of Section 6. Namely, if the man in the carriage covers the distance w in a unit of time—*measured from the train*—then this distance—*as measured from the embankment*—is not necessarily also equal to w.

THE LORENTZ
TRANSFORMATION

The results of the last three sections show that the apparent incompatibility of the law of propagation of light with the principle of relativity (Section 7) has been derived by means of a consideration which borrowed two unjustifiable hypotheses from classical mechanics; these are as follows:

(1) The time-interval (time) between two events is independent of the condition of motion of the body of reference.

(2) The space-interval (distance) between two points of a rigid body is independent of the condition of motion of the body of reference.

If we drop these hypotheses, then the dilemma of Section 7 disappears, because the theorem of the addition of velocities derived in Section 6 becomes invalid. The possibility presents itself that the law of the propagation of light *in vacuo* may be compatible with the principle of relativity, and the question arises: How have we to modify the considerations of Section 6 in order to remove the apparent disagreement between these two fundamental results of experience? This question leads to a general one. In the discussion of Section 6 we have to do with places and times relative both to the train and to the embankment. How are we to find the place and time of an event in relation to the train, when we know the place and time of the event with respect to the railway embankment? Is there a thinkable answer to this question of such a nature that the law of transmission of light *in vacuo* does not contradict the principle of relativity? In other words: Can we conceive of a relation between place and time of the individual events relative to both reference-bodies, such that every ray of light possesses the velocity of transmission *c* relative to the embankment and relative to the train? This question leads to a quite definite positive answer, and to a perfectly definite transformation law for the space-time magnitudes of an event when changing over from one body of reference to another.

Before we deal with this, we shall introduce the following incidental consideration. Up to the present we have only considered events taking place along the embankment, which had mathematically to assume the function of a straight line. In the manner indicated in Section 2 we can imagine this reference-body supplemented laterally and in a vertical direction by means of a framework of rods, so that an event which takes place anywhere can be localised with reference to this framework. Similarly, we can imagine the train travelling with the velocity v to be continued across the whole of space, so that every event, no matter how far off it may be, could also be localised with respect to the second framework. Without committing any fundamental error, we can disregard the fact that in reality these frameworks would continually interfere with each other, owing to the impenetrability of solid bodies. In every such framework we imagine three surfaces perpendicular to each other marked out, and designated as "co-ordinate planes" ("co-ordinate system"). A co-ordinate system K then corresponds to the embankment, and a co-ordinate system K' to the train. An event, wherever it may have taken place, would be fixed in space with respect to K by the three perpendiculars x, y, z on the co-ordinate planes, and with regard to time by a time-value t. Relative to K', *the same event* would be fixed in respect of space and time by corresponding values x', y', z', t', which of course are not identical with x, y, z, t. It has already been set forth in detail how these magnitudes are to be regarded as results of physical measurements.

Obviously our problem can be exactly formulated in the following manner. What are the values x', y', z', t' of an event with respect to K', when the magnitudes x, y, z, t, of the same event with respect to K are given? The relations must be so chosen that the law of the transmission of light *in vacuo* is satisfied for one and the same ray of light (and of course for every ray) with respect to K and K'. For the relative orientation in space of the co-ordinate systems indicated in the diagram (Fig. 2), this problem is solved by means of the equations:

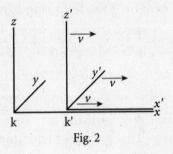

Fig. 2

$$x' = \frac{x - vt}{\sqrt{1 - \dfrac{v^2}{c^2}}}$$

$$y' = y$$
$$z' = z$$

$$t' = \frac{t - \dfrac{v}{c^2} \cdot x}{\sqrt{1 - \dfrac{v^2}{c^2}}}$$

This system of equations is known as the "Lorentz transformation."[1]

If in place of the law of transmission of light we had taken as our basis the tacit assumptions of the older mechanics as to the absolute character of times and lengths, then instead of the above we should have obtained the following equations:

$$x' = x - vt$$
$$y' = y$$
$$z' = z$$
$$t' = t.$$

This system of equations is often termed the "Galilei transformation." The Galilei transformation can be obtained from the Lorentz transformation by substituting an infinitely large value for the velocity of light c in the latter transformation.

Aided by the following illustration, we can readily see that, in accordance with the Lorentz transformation, the law of the transmission of light *in vacuo* is satisfied both for the reference-body K and for the reference-body K'. A light-signal is sent along the positive x-axis,

1. A simple derivation of the Lorentz transformation is given in Appendix 1.

and this light-stimulus advances in accordance with the equation

$$x = ct,$$

i.e. with the velocity *c*. According to the equations of the Lorentz transformation, this simple relation between *x* and *t* involves a relation between *x'* and *t'*. In point of fact, if we substitute for *x* the value *ct* in the first and fourth equations of the Lorentz transformation, we obtain:

$$x' = \frac{(c-v)t}{\sqrt{1-\dfrac{v^2}{c^2}}}$$

$$t' = \frac{\left(1-\dfrac{v}{c}\right)t}{\sqrt{1-\dfrac{v^2}{c^2}}},$$

from which, by division, the expression

$$x' = ct'$$

immediately follows. If referred to the system K', the propagation of light takes place according to this equation. We thus see that the velocity of transmission relative to the reference-body K' is also equal to c. The same result is obtained for rays of light advancing in any other direction whatsoever. Of course this is not surprising, since the equations of the Lorentz transformation were derived conformably to this point of view.

THE BEHAVIOUR OF MEASURING-RODS AND CLOCKS IN MOTION†

I place a metre-rod in the x'-axis of K' in such a manner that one end (the beginning) coincides with the point $x' = 0$, whilst the other end (the end of the rod) coincides with the point $x'= 1$. What is the length of the metre-rod relatively to the system K? In order to learn this, we need only ask where the beginning of the rod and the end of the rod lie with respect to K at a particular time t of the system K. By means of the first equation of the Lorentz transformation the values of these two points at the time $t = 0$ can be shown to be

$$x_{\text{(beginning of rod)}} = 0.\sqrt{1 - \frac{v^2}{c^2}}$$

$$x_{\text{(end of rod)}} = 1.\sqrt{1 - \frac{v^2}{c^2}},$$

the distance between the points being $\sqrt{1 - v^2 / c^2}$.

But the metre-rod is moving with the velocity v relative to K. It therefore follows that the length of a rigid metre-rod moving in the direction of its length with a velocity v is $\sqrt{1 - v^2 / c^2}$ of a metre. The rigid rod is thus shorter when in motion than when at rest, and the more quickly it is moving, the shorter is the rod. For the velocity $v = c$ we should have $\sqrt{1 - v^2 / c^2} = 0$, and for still greater velocities the square-root becomes imaginary. From this we conclude that in the theory of relativity the velocity c plays the part of a limiting velocity, which can neither be reached nor exceeded by any real body.

Of course this feature of the velocity c as a limiting velocity also clearly follows from the equations of the Lorentz transformation, for these become meaningless if we choose values of v greater than c.

If, on the contrary, we had considered a metre-rod at rest in the x-axis with respect to K, then we should have found that the length of the rod as judged from K' would have been $\sqrt{1 - v^2 / c^2}$; this is quite in accordance with the principle of relativity which forms the basis of our considerations.

A priori it is quite clear that we must be able to learn something about the physical behaviour of measuring-rods and clocks from the equations of transformation, for the magnitudes *x, y, z, t*, are nothing more nor less than the results of measurements obtainable by means of measuring-rods and clocks. If we had based our considerations on the Galilei transformation we should not have obtained a contraction of the rod as a consequence of its motion.

Let us now consider a seconds-clock which is permanently situated at the origin ($x' = 0$) of K'. $t' = 0$ and $t' = 1$ are two successive ticks of this clock. The first and fourth equations of the Lorentz transformation give for these two ticks:

$$t = 0$$

and

$$t = \frac{1}{\sqrt{1 - \dfrac{v^2}{c^2}}}.$$

As judged from *K*, the clock is moving with the velocity *v*; as judged from this reference-body, the time which elapses between two strokes of the clock is not one second, but $1/\sqrt{1-v^2/c^2}$ seconds, *i.e.* a somewhat larger time. As a consequence of its motion the clock goes more slowly than when at rest. Here also the velocity *c* plays the part of an unattainable limiting velocity.

THEOREM OF THE ADDITION OF VELOCITIES. THE EXPERIMENT OF FIZEAU

Now in practice we can move clocks and measuring-rods only with velocities that are small compared with the velocity of light; hence we shall hardly be able to compare the results of the previous section directly with the reality. But, on the other hand, these results must strike you as being very singular, and for that reason I shall now draw another conclusion from the theory, one which can easily be derived from the foregoing considerations, and which has been most elegantly confirmed by experiment.

In Section 6 we derived the theorem of the addition of velocities in one direction in the form which also results from the hypotheses of classical mechanics. This theorem can also be deduced readily from the Galilei transformation (Section 11). In place of the man walking inside the carriage, we introduce a point moving relatively to the co-ordinate system K' in accordance with the equation

$$x' = wt'.$$

By means of the first and fourth equations of the Galilei transformation we can express x' and t' in terms of x and t, and we then obtain

$$x = (v + w)t.$$

This equation expresses nothing else than the law of motion of the point with reference to the system K (of the man with reference to the embankment). We denote this velocity by the symbol W, and we then obtain, as in Section 6,

$$W = v + w \dots \dots (A).$$

But we can carry out this consideration just as well on the basis of the theory of relativity. In the equation

$$x' = wt'$$

we must then express x' and t' in terms of x and t, making use of the first and fourth equations of the *Lorentz transformation*. Instead of the equation (A) we then obtain the equation

$$W = \frac{v + w}{1 + \dfrac{vw}{c^2}} \dots\dots(B),$$

which corresponds to the theorem of addition for velocities in one direction according to the theory of relativity. The question now arises as to which of these two theorems is the better in accord with experience. On this point we are enlightened by a most important experiment which the brilliant physicist Fizeau performed more than half a century ago, and which has been repeated since then by some of the best experimental physicists, so that there can be no doubt about its result. The experiment is concerned with the following question. Light travels in a motionless liquid with a particular velocity w. How quickly does it travel in the direction of the arrow in the tube T (see the accompanying diagram, Fig. 3) when the liquid above mentioned is flowing through the tube with a velocity v?

In accordance with the principle of relativity we shall certainly have to take for granted that the propagation of light always takes place with the same velocity w *with respect to the liquid*, whether the latter is in motion with reference to other bodies or not. The velocity of light relative to the liquid and the velocity of the latter relative to the tube are thus known, and we require the velocity of light relative to the tube.

It is clear that we have the problem of Section 6 again before us. The tube plays the part of

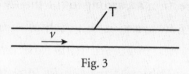

Fig. 3

the railway embankment or of the co-ordinate system *K*, the liquid plays the part of the carriage or of the co-ordinate system *K'*, and finally, the light plays the part of the man walking along the carriage, or of the moving point in the present section. If we denote the velocity of the light relative to the tube by *W*, then this is given by the equation (A) or (B), according as the Galilei transformation or the Lorentz transformation corresponds to the facts. Experiment[1] decides in favour of equation (B) derived from the theory of relativity, and the agreement is, indeed, very exact. According to recent and most excellent measurements by Zeeman, the influence of the

---

1. Fizeau found $W = w + v\left(1 - 1/n^2\right)$ where $n = c/w$ the index of refraction of the liquid. On the other hand, owing to the smallness of vw/c^2 as compared with 1, we can replace (B) in the first place by $W = (w + v)\left(1 - vw/c^2\right)$ or to the same order of approximation by $w + v\left(1 - 1/n^2\right)$ which agrees with Fizeau's result.

velocity of flow v on the propagation of light is represented by formula (B) to within one per cent.

Nevertheless we must now draw attention to the fact that a theory of this phenomenon was given by H. A. Lorentz long before the statement of the theory of relativity. This theory was of a purely electrodynamical nature, and was obtained by the use of particular hypotheses as to the electromagnetic structure of matter. This circumstance, however, does not in the least diminish the conclusiveness of the experiment as a crucial test in favour of the theory of relativity, for the electrodynamics of Maxwell-Lorentz, on which the original theory was based, in no way opposes the theory of relativity. Rather has the latter been developed from electrodynamics as an astoundingly simple combination and generalisation of the hypotheses, formerly independent of each other, on which electrodynamics was built.

THE HEURISTIC VALUE OF THE THEORY OF RELATIVITY

Our train of thought in the foregoing pages can be epitomised in the following manner. Experience has led to the conviction that, on the one hand, the principle of relativity holds true, and that on the other hand the velocity of transmission of light *in vacuo* has to be considered equal to a constant *c*. By uniting these two postulates we obtained the law of transformation for the rectangular co-ordinates *x, y, z* and the time *t* of the events which constitute the processes of nature. In this connection we did not obtain the Galilei transformation, but, differing from classical mechanics, the *Lorentz transformation*.

The law of transmission of light, the acceptance of which is justified by our actual knowledge, played an important part in this process of thought. Once in possession of the Lorentz transformation, however, we can combine this with the principle of relativity, and sum up the theory thus:

Every general law of nature must be so constituted that it is transformed into a law of exactly the same form when, instead of the space-time variables x, y, z, t of the original co-ordinate system K, we introduce new space-time variables x', y', z', t' of a co-ordinate system K'. In this connection the relation between the ordinary and the accented magnitudes is given by the Lorentz transformation. Or, in brief: General laws of nature are co-variant with respect to Lorentz transformations.

This is a definite mathematical condition that the theory of relativity demands of a natural law, and in virtue of this, the theory becomes a valuable heuristic aid in the search for general laws of nature. If a general law of nature were to be found which did not satisfy this condition, then at least one of the two fundamental assumptions of the theory would have been disproved. Let us now examine what general results the latter theory has hitherto evinced.

15

GENERAL RESULTS
OF THE THEORY†

I t is clear from our previous considerations that the (special) theory of relativity has grown out of electrodynamics and optics. In these fields it has not appreciably altered the predictions of theory, but it has considerably simplified the theoretical structure, *i.e.* the derivation of laws, and—what is incomparably more important—it has considerably reduced the number of independent hypotheses forming the basis of theory. The special theory of relativity has rendered the Maxwell-Lorentz theory so plausible, that the latter would have been generally accepted by physicists even if experiment had decided less unequivocally in its favour.

Classical mechanics required to be modified before it could come into line with the demands of the special theory of relativity. For the main part, however, this modification affects only the laws for rapid motions, in which the velocities of matter v are not very small as compared with the velocity of light. We have experience of such rapid motions only in the case of electrons and ions; for other motions the variations from the laws of classical mechanics are too small to make themselves evident in practice. We shall not consider the motion of stars until we come to speak of the general theory of relativity. In accordance with the theory of relativity the kinetic energy of a material point of mass m is no longer given by the well-known expression

$$m\frac{v^2}{2},$$

but by the expression

$$\frac{mc^2}{\sqrt{1-\dfrac{v^2}{c^2}}}.$$

This expression approaches infinity as the velocity v approaches the velocity of light c. The velocity must therefore always remain less than c, however great may be the energies used to produce the acceleration. If we develop the expression for the kinetic energy in the form of a series, we obtain

$$mc^2 + m\frac{v^2}{2} + \frac{3}{8}m\frac{v^4}{c^2} + \dots.$$

When v^2/c^2 is small compared with unity, the third of these terms is always small in comparison with the second, which last is alone considered in classical mechanics. The first term mc^2 does not contain the velocity, and requires no consideration if we are only dealing with the question as to how the energy of a point-mass depends on the velocity. We shall speak of its essential significance later.

The most important result of a general character to which the special theory of relativity has led is concerned with the conception of mass. Before the advent of relativity, physics recognised two conservation laws of fundamental importance, namely, the law of the conservation of energy and the law of the conservation of mass; these two fundamental laws appeared to be quite independent of each other. By means of the theory of relativity they have been united into one law. We shall now briefly consider how this unification came about, and what meaning is to be attached to it.

The principle of relativity requires that the law of the conservation of energy should hold not only with reference to a co-ordinate system K, but also with respect to every co-ordinate system K' which is in a state of uniform motion of translation relative to K, or, briefly, relative to every "Galileian" system of co-ordinates. In contrast to classical mechanics, the Lorentz transformation is the deciding factor in the transition from one such system to another.

By means of comparatively simple considerations we are led to draw the following conclusion from these premises, in conjunction with the fundamental equations of the electrodynamics of Maxwell: A body moving with the velocity v, which absorbs[1] an amount of energy E_0 in the form of radiation without suffering an alteration in velocity in the process, has, as a consequence, its energy increased by an amount

$$\frac{E_0}{\sqrt{1 - \dfrac{v^2}{c^2}}}.$$

In consideration of the expression given above for the kinetic energy of the body, the required energy of the body comes out to be

$$\frac{\left(m + \dfrac{E_0}{c^2}\right)c^2}{\sqrt{1 - \dfrac{v^2}{c^2}}}.$$

1. E_0 is the energy taken up, as judged from a co-ordinate system moving with the body.

Thus the body has the same energy as a body of mass

$$\left(m + \frac{E_0}{c^2}\right)$$

moving with the velocity v. Hence we can say: If a body takes up an amount of energy E_0, then its inertial mass increases by an amount E_0/c^2; the inertial mass of a body is not a constant, but varies according to the change in the energy of the body. The inertial mass of a system of bodies can even be regarded as a measure of its energy. The law of the conservation of the mass of a system becomes identical with the law of the conservation of energy, and is only valid provided that the system neither takes up nor sends out energy. Writing the expression for the energy in the form

$$\frac{mc^2 + E_0}{\sqrt{1 - \frac{v^2}{c^2}}},$$

we see that the term mc^2, which has hitherto attracted our attention, is nothing else than the energy possessed by the body[2] before it absorbed the energy E_0.

A direct comparison of this relation with experiment is not possible at the present time, owing to the fact that the changes in energy E_0 to which we can subject a system are not large enough to make themselves perceptible as a change in the inertial mass of the system. E_0/c^2 is too small in comparison with the mass m, which was present before the alteration of the energy. It is owing to this circumstance that classical mechanics was able to establish successfully the conservation of mass as a law of independent validity.

Let me add a final remark of a fundamental nature. The success of the Faraday-Maxwell interpretation of electromagnetic action at a distance resulted in physicists becoming convinced that there are no such things as instantaneous actions at a distance (not involving an intermediary medium) of the type of Newton's law of gravitation. According to the theory of relativity, action at a distance with the velocity of light always takes the place of instantaneous action at a distance or of action

2. As judged from a co-ordinate system moving with the body.

at a distance with an infinite velocity of transmission. This is connected with the fact that the velocity c plays a fundamental rôle in this theory. In Part II we shall see in what way this result becomes modified in the general theory of relativity.

16

EXPERIENCE AND THE SPECIAL THEORY OF RELATIVITY

To what extent is the special theory of relativity supported by experience? This question is not easily answered for the reason already mentioned in connection with the fundamental experiment of Fizeau. The special theory of relativity has crystallised out from the Maxwell-Lorentz theory of electromagnetic phenomena. Thus all facts of experience which support the electromagnetic theory also support the theory of relativity. As being of particular importance, I mention here the fact that the theory of relativity enables us to predict the effects produced on the light reaching us from the fixed stars. These results are obtained in an exceedingly simple manner, and the effects indicated, which are due to the relative motion of the earth with reference to those fixed stars, are found to be in accord with experience. We refer to the yearly movement of the apparent position of the fixed stars resulting from the motion of the earth round the sun (aberration), and to the influence of the radial components of the relative motions of the fixed stars with respect to the earth on the colour of the light reaching us from them. The latter effect manifests itself in a slight displacement of the spectral lines of the light transmitted to us from a fixed star, as compared with the position of the same spectral lines when they are produced by a terrestrial source of light (Doppler principle). The experimental arguments in favour of the Maxwell-Lorentz theory, which are at the same time arguments in favour of the theory of relativity, are too numerous to be set forth here. In reality they limit the theoretical possibilities to such an extent, that no other theory than that of Maxwell and Lorentz has been able to hold its own when tested by experience.

But there are two classes of experimental facts hitherto obtained which can be represented in the Maxwell-Lorentz theory only by the introduction of an auxiliary hypothesis, which in itself—*i.e.* without making use of the theory of relativity—appears extraneous.

It is known that cathode rays and the so-called β-rays emitted by radioactive substances consist of negatively electrified particles (electrons) of very small inertia and large velocity. By examining the deflection of these rays under the influence of electric and magnetic fields, we can study the law of motion of these particles very exactly.

In the theoretical treatment of these electrons, we are faced with the difficulty that electrodynamic theory of itself is unable to give an account of their nature. For since electrical masses of one sign repel each other, the negative electrical masses constituting the electron would necessarily be scattered under the influence of their mutual repulsions, unless there are forces of another kind operating between them, the nature of which has hitherto remained obscure to us.[1] If we now assume that the relative distances between the electrical masses constituting the electron remain unchanged during the motion of the electron (rigid connection in the sense of classical mechanics), we arrive at a law of motion of the electron which does not agree with experience. Guided by purely formal points of view, H. A. Lorentz was the first to introduce the hypothesis that the particles constituting the electron experience a contraction in the direction of motion in consequence of that motion, the amount of this contraction being proportional to the expression $\sqrt{1 - v^2 / c^2}$.

---

1. The general theory of relativity renders it likely that the electrical masses of an electron are held together by gravitational forces.

This hypothesis, which is not justifiable by any electrodynamical facts, supplies us then with that particular law of motion which has been confirmed with great precision in recent years.

The theory of relativity leads to the same law of motion, without requiring any special hypothesis whatsoever as to the structure and the behaviour of the electron. We arrived at a similar conclusion in Section 13 in connection with the experiment of Fizeau, the result of which is foretold by the theory of relativity without the necessity of drawing on hypotheses as to the physical nature of the liquid.

The second class of facts to which we have alluded has reference to the question whether or not the motion of the earth in space can be made perceptible in terrestrial experiments. We have already remarked in Section 5 that all attempts of this nature led to a negative result. Before the theory of relativity was put forward, it was difficult to become reconciled to this negative result, for reasons now to be discussed. The inherited prejudices about time and space did not allow any doubt to arise as to the prime importance of the Galilei transformation for changing over from one body of reference to another. Now assuming that the Maxwell-Lorentz equations hold for a reference-body K, we then find that they do not hold for a reference-body K' moving uniformly with respect to K, if we assume that the relations of the Galileian transformation exist between the co-ordinates of K and K'. It thus appears that of all Galileian co-ordinate systems one (K) corresponding to a particular state of motion is physically unique. This result was interpreted physically by regarding K as at rest with respect to a hypothetical æther of space. On the other hand, all co-ordinate systems K' moving relatively to K were to be regarded as in motion with respect to the æther. To this motion of K' against the æther ("æther-drift" relative to K') were assigned the more complicated laws which were supposed to hold relative to K'. Strictly speaking, such an æther-drift ought also to be assumed relative to the earth, and for a long time the efforts of physicists were devoted to attempts to detect the existence of an æther-drift at the earth's surface.

In one of the most notable of these attempts Michelson devised a method which appears as though it must be decisive. Imagine two mirrors so arranged on a rigid body that the reflecting surfaces face each other. A ray of light requires a perfectly definite time T to pass from one mirror to the other and back again, if the whole system be at rest with respect to the æther. It is found by calculation, however, that a slightly different time T' is required for this process, if the body, together with the mirrors, be moving relatively to the æther. And yet another point: it is shown by calculation that for a given velocity v with reference to the æther, this time T' is different when the body is moving perpendicularly to the planes of the mirrors from that resulting when the motion is parallel to these planes. Although the estimated difference between these two times is exceedingly small, Michelson and Morley performed an experiment involving interference in which this difference should have been clearly detectable. But the experiment gave a negative result—a fact very perplexing to physicists. Lorentz and FitzGerald rescued the theory from this difficulty by assuming that the motion of the body relative to the æther produces a contraction of the body in the direction of motion, the amount of contraction being just sufficient to compensate for the difference in time mentioned above. Comparison with the discussion in Section 12 shows that from the standpoint also of the theory of relativity this solution of the difficulty was the right one. But on the basis of the theory of relativity the method of interpretation is incomparably more satisfactory. According to this theory there is no such thing as a "specially favoured" (unique) co-ordinate system to occasion the introduction of the æther-idea, and hence there can be no æther-drift, nor any experiment with which to demonstrate it. Here the contraction of moving bodies follows from the two fundamental principles of the theory without the introduction of particular hypotheses; and as the prime factor involved in this contraction we find, not the motion in itself, to which we cannot attach any meaning, but the motion with respect to the body of reference chosen in the particular case in point. Thus for a co-ordinate system moving with the earth the mirror system of Michelson and Morley is not shortened, but it *is* shortened for a co-ordinate system which is at rest relatively to the sun.

MINKOWSKI'S
FOUR-DIMENSIONAL SPACE†

The non-mathematician is seized by a mysterious shuddering when he hears of "four-dimensional" things, by a feeling not unlike that awakened by thoughts of the occult. And yet there is no more common-place statement than that the world in which we live is a four-dimensional space-time continuum.

Space is a three-dimensional continuum. By this we mean that it is possible to describe the position of a point (at rest) by means of three numbers (co-ordinates) x, y, z, and that there is an indefinite number of points in the neighbourhood of this one, the position of which can be described by co-ordinates such as x_1, y_1, z_1, which may be as near as we choose to the respective values of the co-ordinates x, y, z of the first point. In virtue of the latter property we speak of a "continuum," and owing to the fact that there are three co-ordinates we speak of it as being "three-dimensional."

Similarly, the world of physical phenomena which was briefly called "world" by Minkowski is naturally four-dimensional in the space-time sense. For it is composed of individual events, each of which is described by four numbers, namely, three space co-ordinates x, y, z and a time co-ordinate, the time-value t. The "world" is in this sense also a continuum; for to every event there are as many "neighbouring" events (realised or at least thinkable) as we care to choose, the co-ordinates x_1, y_1, z_1, t_1 of which differ by an indefinitely small amount from those of the event x, y, z, t originally considered. That we have not been accustomed to regard the world in this sense as a four-dimensional continuum is due to the fact that in physics, before the advent of the theory of relativity, time played a different and more independent rôle, as compared with the space co-ordinates. It is for this reason that we have been in the habit of treating time as an independent continuum. As a matter of fact, according to classical mechanics, time is absolute, *i.e.* it is independent of the position and the condition of motion of the system of co-ordinates. We see this expressed in the last equation of the Galileian transformation ($t' = t$).

The four-dimensional mode of consideration of the "world" is natural on the theory of relativity, since according to this theory time is robbed of its independence. This is shown by the fourth equation of the Lorentz transformation:

$$t' = \frac{t - \dfrac{v}{c^2}x}{\sqrt{1 - \dfrac{v^2}{c^2}}}.$$

Moreover, according to this equation the time difference $\Delta t'$ of two events with respect to K' does not in general vanish, even when the time difference Δt of the same events with reference to K vanishes. Pure "space-distance" of two events with respect to K results in "time-distance" of the same events with respect to K'. But the discovery, of Minkowski, which was of importance for the formal development of the theory of relativity, does not lie here. It is to be found rather in the fact of his recognition that the four-dimensional space-time continuum of the theory of relativity, in its most essential formal properties, shows a pronounced relationship to the three-dimensional continuum of Euclidean geometrical space.[1] In order to give due prominence to this relationship, however, we must replace the usual time co-ordinate t by an imaginary

1. Cf. the somewhat more detailed discussion in Appendix 2.

magnitude $\sqrt{-1} \cdot ct$ proportional to it. Under these conditions, the natural laws satisfying the demands of the (special) theory of relativity assume mathematical forms, in which the time co-ordinate plays exactly the same rôle as the three space co-ordinates. Formally, these four co-ordinates correspond exactly to the three space co-ordinates in Euclidean geometry. It must be clear even to the non-mathematician that, as a consequence of this purely formal addition to our knowledge, the theory perforce gained clearness in no mean measure.

These inadequate remarks can give the reader only a vague notion of the important idea contributed by Minkowski. Without it the general theory of relativity, of which the fundamental ideas are developed in the following pages, would perhaps have got no farther than its long clothes. Minkowski's work is doubtless difficult of access to anyone inexperienced in mathematics, but since it is not necessary to have a very exact grasp of this work in order to understand the fundamental ideas of either the special or the general theory of relativity, I shall at present leave it here, and shall revert to it only towards the end of Part II.

PART II

THE GENERAL THEORY OF RELATIVITY

18

SPECIAL AND GENERAL PRINCIPLE OF RELATIVITY

The basal principle, which was the pivot of all our previous considerations, was the *special* principle of relativity, *i.e.* the principle of the physical relativity of all *uniform* motion. Let us once more analyse its meaning carefully.

It was at all times clear that, from the point of view of the idea it conveys to us, every motion must only be considered as a relative motion. Returning to the illustration we have frequently used of the embankment and the railway carriage, we can express the fact of the motion here taking place in the following two forms, both of which are equally justifiable:

77

(*a*) The carriage is in motion relative to the
embankment.

(*b*) The embankment is in motion relative to the
carriage.

In (*a*) the embankment, in (*b*) the carriage, serves as
the body of reference in our statement of the motion
taking place. If it is simply a question of detecting or of
describing the motion involved, it is in principle immaterial to what reference-body we refer the motion. As
already mentioned, this is self-evident, but it must not
be confused with the much more comprehensive statement called "the principle of relativity," which we have
taken as the basis of our investigations.

The principle we have made use of not only main-
tains that we may equally well choose the carriage or the
embankment as our reference-body for the description
of any event (for this, too, is self-evident). Our principle
rather asserts what follows: If we formulate the general
laws of nature as they are obtained from experience, by
making use of

(*a*) the embankment as reference-body,

(*b*) the railway carriage as reference-body,

then these general laws of nature (*e.g.* the laws of
mechanics or the law of the propagation of light *in*

vacuo) have exactly the same form in both cases. This can also be expressed as follows: For the *physical* description of natural processes, neither of the reference-bodies *K*, *K'* is unique (lit. "specially marked out") as compared with the other. Unlike the first, this latter statement need not of necessity hold *a priori*; it is not contained in the conceptions of "motion" and "reference-body" and derivable from them; only *experience* can decide as to its correctness or incorrectness.

Up to the present, however, we have by no means maintained the equivalence of *all* bodies of reference *K* in connection with the formulation of natural laws. Our course was more on the following lines. In the first place, we started out from the assumption that there exists a reference-body *K*, whose condition of motion is such that the Galileian law holds with respect to it: A particle left to itself and sufficiently far removed from all other particles moves uniformly in a straight line. With reference to *K* (Galileian reference-body) the laws of nature were to be as simple as possible. But in addition to *K*, all bodies of reference *K'* should be given preference in this sense, and they should be exactly equivalent to *K* for the formulation of natural laws, provided that they are in a state of *uniform rectilinear and non-rotary motion* with respect to *K*; all these bodies of reference are to be regarded as Galileian reference-bodies. The validity of the principle of relativity was assumed only for these reference-bodies, but not for others (*e.g.* those possessing motion of a different kind). In this sense we speak of the *special* principle of relativity, or special theory of relativity.

In contrast to this we wish to understand by the "general principle of relativity" the following statement: All bodies of reference K, K', etc., are equivalent for the description of natural phenomena (formulation of the general laws of nature), whatever may be their state of motion. But before proceeding farther, it ought to be pointed out that this formulation must be replaced later by a more abstract one, for reasons which will become evident at a later stage.

Since the introduction of the special principle of relativity has been justified, every intellect which strives after generalisation must feel the temptation to venture the step towards the general principle of relativity. But a simple and apparently quite reliable consideration seems to suggest that, for the present at any rate, there is little hope of success in such an attempt. Let us imagine ourselves transferred to our old friend the railway carriage, which is travelling at a uniform rate. As long as it is moving uniformly, the occupant of the carriage is not sensible of its motion, and it is for this reason that he can unreluctantly interpret the facts of the case as indicating that the carriage is at rest, but the embankment in motion. Moreover, according to the special principle of relativity, this interpretation is quite justified also from a physical point of view.

If the motion of the carriage is now changed into a non-uniform motion, as for instance by a powerful application of the brakes, then the occupant of the carriage experiences a correspondingly powerful jerk forwards. The retarded motion is manifested in the mechanical behaviour of bodies relative to the person in the railway carriage. The mechanical behaviour is different from that of the case previously considered, and for this reason it would appear to be impossible that the same mechanical laws hold relatively to the non-uniformly moving carriage, as hold with reference to the carriage when at rest or in uniform motion. At all events it is clear that the Galileian law does not hold with respect to the non-uniformly moving carriage. Because of this, we feel compelled at the present juncture to grant a kind of absolute physical reality to non-uniform motion, in opposition to the general principle of relativity. But in what follows we shall soon see that this conclusion cannot be maintained.

19

THE GRAVITATIONAL FIELD†

If we pick up a stone and then let it go, why does it fall to the ground?" The usual answer to this question is: "Because it is attracted by the earth." Modern physics formulates the answer rather different-ly for the following reason. As a result of the more care-ful study of electromagnetic phenomena, we have come to regard action at a distance as a process impossible without the intervention of some intermediary medi-um. If, for instance, a magnet attracts a piece of iron, we cannot be content to regard this as meaning that the magnet acts directly on the iron through the intermedi-ate empty space, but we are constrained to imagine—after the manner of Faraday—that the magnet always calls into being something physically real in the space around it, that something being what we call a "magnet-ic field." In its turn this magnetic field operates on the piece of iron, so that the latter strives to move towards the magnet. We shall not discuss here the justification for this incidental conception, which is indeed a some-what arbitrary one. We shall only mention that with its aid electromagnetic phenomena can be theoretically represented much more satisfactorily than without it, and this applies particularly to the transmission of electromagnetic waves. The effects of gravitation also are regarded in an analogous manner.

The action of the earth on the stone takes place indirectly. The earth produces in its surroundings a gravitational field, which acts on the stone and produces its motion of fall. As we know from experience, the intensity of the action on a body diminishes according to a quite definite law, as we proceed farther and farther away from the earth. From our point of view this means: The law governing the properties of the gravitational field in space must be a perfectly definite one, in order correctly to represent the diminution of gravitational action with the distance from operative bodies. It is something like this: The body (*e.g.* the earth) produces a field in its immediate neighbourhood directly; the intensity and direction of the field at points farther removed from the body are thence determined by the law which governs the properties in space of the gravitational fields themselves.

In contrast to electric and magnetic fields, the gravitational field exhibits a most remarkable property, which is of fundamental importance for what follows. Bodies which are moving under the sole influence of a gravitational field receive an acceleration, *which does not in the least depend either on the material or on the physical state of the body.* For instance, a piece of lead and a piece of wood fall in exactly the same manner in a gravitational field (*in vacuo*), when they start off from rest or with the same initial velocity. This law, which holds most accurately, can be expressed in a different form in the light of the following consideration.

According to Newton's law of motion, we have

$$\text{(Force)} = \text{(inertial mass)} \times \text{(acceleration)},$$

where the "inertial mass" is a characteristic constant of the accelerated body. If now gravitation is the cause of the acceleration, we then have

$$\text{(Force)} = \text{(gravitational mass)} \times \text{(intensity of the gravitational field)},$$

where the "gravitational mass" is likewise a characteristic constant for the body. From these two relations follows:

$$\text{(acceleration)} = \frac{\text{(gravitational mass)}}{\text{(inertial mass)}} \times \text{(intensity of the gravitational field)}.$$

If now, as we find from experience, the acceleration is to be independent of the nature and the condition of the body and always the same for a given gravitational field, then the ratio of the gravitational to the inertial mass must likewise be the same for all bodies. By a suitable choice of units we can thus make this ratio equal to unity. We then have the following law: The *gravitational* mass of a body is equal to its *inertial* mass.

It is true that this important law had hitherto been recorded in mechanics, but it had not been *interpreted*. A satisfactory interpretation can be obtained only if we recognise the following fact: *The same* quality of a body manifests itself according to circumstances as "inertia" or as "weight" (lit. "heaviness"). In the following section we shall show to what extent this is actually the case, and how this question is connected with the general postulate of relativity.

20

THE EQUALITY OF INERTIAL AND GRAVITATIONAL MASS AS AN ARGUMENT FOR THE GENERAL POSTULATE OF RELATIVITY

We imagine a large portion of empty space, so far removed from stars and other appreciable masses that we have before us approximately the conditions required by the fundamental law of Galilei. It is then possible to choose a Galileian reference-body for this part of space (world), relative to which points at rest remain at rest and points in motion continue permanently in uniform rectilinear motion. As reference-body let us imagine a spacious chest resembling a room with an observer inside who is equipped with apparatus. Gravitation naturally does not exist for this observer. He must fasten himself with strings to the floor, otherwise the slightest impact against the floor will cause him to rise slowly towards the ceiling of the room.

To the middle of the lid of the chest is fixed externally a hook with rope attached, and now a "being" (what kind of a being is immaterial to us) begins pulling at this with a constant force. The chest together with the observer then begin to move "upwards" with a uniformly accelerated motion. In course of time their velocity will reach unheard-of values—provided that we are viewing all this from another reference-body which is not being pulled with a rope.

But how does the man in the chest regard the process? The acceleration of the chest will be transmitted to him by the reaction of the floor of the chest. He must therefore take up this pressure by means of his legs if he does not wish to be laid out full length on the floor. He is then standing in the chest in exactly the same way as anyone stands in a room of a house on our earth. If he releases a body which he previously had in his hand, the acceleration of the chest will no longer be transmitted to this body, and for this reason the body will approach the floor of the chest with an accelerated relative motion. The observer will further convince himself *that the acceleration of the body towards the floor of the chest is always of the same magnitude, whatever kind of body he may happen to use for the experiment.*

Relying on his knowledge of the gravitational field (as it was discussed in the preceding section), the man in the chest will thus come to the conclusion that he and the chest are in a gravitational field which is constant with regard to time. Of course he will be puzzled for a moment as to why the chest does not fall in this gravitational field. Just then, however, he discovers the hook in the middle of the lid of the chest and the rope which is attached to it, and he consequently comes to the conclusion that the chest is suspended at rest in the gravitational field.

Ought we to smile at the man and say that he errs in his conclusion? I do not believe we ought if we wish to remain consistent; we must rather admit that his mode of grasping the situation violates neither reason nor known mechanical laws. Even though it is being accelerated with respect to the "Galileian space" first considered, we can nevertheless regard the chest as being at rest. We have thus good grounds for extending the principle of relativity to include bodies of reference which are accelerated with respect to each other, and as a result we have gained a powerful argument for a generalised postulate of relativity.

We must note carefully that the possibility of this mode of interpretation rests on the fundamental property of the gravitational field of giving all bodies the same acceleration, or, what comes to the same thing, on the law of the equality of inertial and gravitational mass.

If this natural law did not exist, the man in the accelerated chest would not be able to interpret the behaviour of the bodies around him on the supposition of a gravitational field, and he would not be justified on the grounds of experience in supposing his reference-body to be "at rest."

Suppose that the man in the chest fixes a rope to the inner side of the lid, and that he attaches a body to the free end of the rope. The result of this will be to stretch the rope so that it will hang "vertically" downwards. If we ask for an opinion of the cause of tension in the rope, the man in the chest will say: "The suspended body experiences a downward force in the gravitational field, and this is neutralised by the tension of the rope; what determines the magnitude of the tension of the rope is the *gravitational mass* of the suspended body." On the other hand, an observer who is poised freely in space will interpret the condition of things thus: "The rope must perforce take part in the accelerated motion of the chest, and it transmits this motion to the body attached to it. The tension of the rope is just large enough to effect the acceleration of the body. That which determines the magnitude of the tension of the rope is the *inertial mass* of the body." Guided by this example, we see that our extension of the principle of relativity

implies the *necessity* of the law of the equality of inertial
and gravitational mass. Thus we have obtained a physical interpretation of this law.

From our consideration of the accelerated chest we
see that a general theory of relativity must yield important results on the laws of gravitation. In point of fact,
the systematic pursuit of the general idea of relativity
has supplied the laws satisfied by the gravitational field.
Before proceeding farther, however, I must warn the
reader against a misconception suggested by these considerations. A gravitational field exists for the man in
the chest, despite the fact that there was no such field for
the co-ordinate system first chosen. Now we might easily suppose that the existence of a gravitational field is
always only an *apparent* one. We might also think that,
regardless of the kind of gravitational field which may
be present, we could always choose another reference-
body such that *no* gravitational field exists with reference to it. This is by no means true for all gravitational
fields, but only for those of quite special form. It is, for
instance, impossible to choose a body of reference such
that, as judged from it, the gravitational field of the
earth (in its entirety) vanishes.

We can now appreciate why that argument is not
convincing, which we brought forward against the gen-

eral principle of relativity at the end of Section 18. It is certainly true that the observer in the railway carriage experiences a jerk forwards as a result of the application of the brake, and that he recognises in this the non-uniformity of motion (retardation) of the carriage. But he is compelled by nobody to refer this jerk to a "real" acceleration (retardation) of the carriage. He might also interpret his experience thus: "My body of reference (the carriage) remains permanently at rest. With reference to it, however, there exists (during the period of application of the brakes) a gravitational field which is directed forwards and which is variable with respect to time. Under the influence of this field, the embankment together with the earth moves non-uniformly in such a manner that their original velocity in the backwards direction is continuously reduced."

21

IN WHAT RESPECTS ARE THE FOUNDATIONS OF CLASSICAL MECHANICS AND OF THE SPECIAL THEORY OF RELATIVITY UNSATISFACTORY?

We have already stated several times that classical mechanics starts out from the following law: Material particles sufficiently far removed from other material particles continue to move uniformly in a straight line or continue in a state of rest. We have also repeatedly emphasised that this fundamental law can only be valid for bodies of reference K which possess certain unique states of motion, and which are in uniform translational motion relative to each other. Relative to other reference-bodies K the law is not valid. Both in classical mechanics and in the special theory of relativity we therefore differentiate between reference-bodies K relative to which the recognised "laws of nature" can be said to hold, and reference-bodies K relative to which these laws do not hold.

But no person whose mode of thought is logical can rest satisfied with this condition of things. He asks: "How does it come that certain reference-bodies (or their states of motion) are given priority over other reference-bodies (or their states of motion)? *What is the reason for this preference?* In order to show clearly what I mean by this question, I shall make use of a comparison.

I am standing in front of a gas range. Standing alongside of each other on the range are two pans so much alike that one may be mistaken for the other. Both are half full of water. I notice that steam is being emitted continuously from the one pan, but not from the other. I am surprised at this, even if I have never seen either a gas range or a pan before. But if I now notice a luminous something of bluish colour under the first pan but not under the other, I cease to be astonished, even if I have never before seen a gas flame. For I can only say that this bluish something will cause the emission of the steam, or at least *possibly* it may do so. If, however, I notice the bluish something in neither case, and if I observe that the one continuously emits steam whilst the other does not, then I shall remain astonished and dissatisfied until I have discovered some circumstance to which I can attribute the different behaviour of the two pans.

Analogously, I seek in vain for a real something in classical mechanics (or in the special theory of relativity) to which I can attribute the different behaviour of bodies considered with respect to the reference-systems K and K'.[1] Newton saw this objection and attempted to invalidate it, but without success. But E. Mach recognised it most clearly of all, and because of this objection he claimed that mechanics must be placed on a new basis. It can only be got rid of by means of a physics which is conformable to the general principle of relativity, since the equations of such a theory hold for every body of reference, whatever may be its state of motion.

---

1. The objection is of importance more especially when the state of motion of the reference-body is of such a nature that it does not require any external agency for its maintenance, *e.g.* in the case when the reference-body is rotating uniformly.

22

A FEW INFERENCES FROM THE GENERAL PRINCIPLE OF RELATIVITY

The considerations of Section 20 show that the general principle of relativity puts us in a position to derive properties of the gravitational field in a purely theoretical manner. Let us suppose, for instance, that we know the space-time "course" for any natural process whatsoever, as regards the manner in which it takes place in the Galileian domain relative to a Galileian body of reference K. By means of purely theoretical operations (*i.e.* simply by calculation) we are then able to find how this known natural process appears, as seen from a reference-body K' which is accelerated relatively to K. But since a gravitational field exists with respect to this new body of reference K', our consideration also teaches us how the gravitational field influences the process studied.

For example, we learn that a body which is in a state of uniform rectilinear motion with respect to K (in accordance with the law of Galilei) is executing an accelerated and in general curvilinear motion with respect to

the accelerated reference-body K' (chest). This acceleration or curvature corresponds to the influence on the
moving body of the gravitational field prevailing relatively to K'. It is known that a gravitational field influences the movement of bodies in this way, so that our
consideration supplies us with nothing essentially new.

However, we obtain a new result of fundamental
importance when we carry out the analogous consideration for a ray of light. With respect to the Galileian
reference-body K, such a ray of light is transmitted rec-
tilinearly with the velocity c. It can easily be shown that
the path of the same ray of light is no longer a straight
line when we consider it with reference to the accelerated chest (reference-body K'). From this we conclude,
*that, in general, rays of light are propagated curvilinearly
in gravitational fields*. In two respects this result is of
great importance.

In the first place, it can be compared with the reality. Although a detailed examination of the question
shows that the curvature of light rays required by the
general theory of relativity is only exceedingly small for
the gravitational fields at our disposal in practice, its
estimated magnitude for light rays passing the sun at
grazing incidence is nevertheless 1.7 seconds of arc. This
ought to manifest itself in the following way. As seen

from the earth, certain fixed stars appear to be in the neighbourhood of the sun, and are thus capable of observation during a total eclipse of the sun. At such times, these stars ought to appear to be displaced outwards from the sun by an amount indicated above, as compared with their apparent position in the sky when the sun is situated at another part of the heavens. The examination of the correctness or otherwise of this deduction is a problem of the greatest importance, the early solution of which is to be expected of astronomers.[1]

In the second place our result shows that, according to the general theory of relativity, the law of the constancy of the velocity of light *in vacuo*, which constitutes one of the two fundamental assumptions in the special theory of relativity and to which we have already frequently referred, cannot claim any unlimited validity. A curvature of rays of light can only take place when the velocity of propagation of light varies with position. Now we might think that as a consequence of this, the special theory of relativity and with it the whole

1. By means of the star photographs of two expeditions equipped by a Joint Committee of the Royal and Royal Astronomical Societies, the existence of the deflection of light demanded by theory was confirmed during the solar eclipse of 29th May, 1919. (Cf. Appendix 3.)

theory of relativity would be laid in the dust. But in reality this is not the case. We can only conclude that the special theory of relativity cannot claim an unlimited domain of validity; its result hold only so long as we are able to disregard the influences of gravitational fields on the phenomena (*e.g.* of light).

Since it has often been contended by opponents of the theory of relativity that the special theory of relativity is overthrown by the general theory of relativity, it is perhaps advisable to make the facts of the case clearer by means of an appropriate comparison. Before the development of electrodynamics the laws of electrostatics were looked upon as the laws of electricity. At the present time we know that electric fields can be derived correctly from electrostatic considerations only for the case, which is never strictly realised, in which the electrical masses are quite at rest relatively to each other, and to the co-ordinate system. Should we be justified in saying that for this reason electrostatics is overthrown by the field-equations of Maxwell in electrodynamics? Not in the least. Electrostatics is contained in electrodynamics as a limiting case; the laws of the latter lead directly to those of the former for the case in which the fields are invariable with regard to time. No fairer destiny could be allotted to any physical theory, than that it should of itself point out the way to the introduction of a more comprehensive theory, in which it lives on as a limiting case.

In the example of the transmission of light just dealt with, we have seen that the general theory of relativity enables us to derive theoretically the influence of a gravitational field on the course of natural processes, the laws of which are already known when a gravitational field is absent. But the most attractive problem, to the solution of which the general theory of relativity supplies the key, concerns the investigation of the laws satisfied by the gravitational field itself. Let us consider this for a moment.

We are acquainted with space-time domains which behave (approximately) in a "Galileian" fashion under suitable choice of reference-body, *i.e.* domains in which gravitational fields are absent. If we now refer such a domain to a reference-body K' possessing any kind of motion, then relative to K' there exists a gravitational field which is variable with respect to space and time.[1] The character of this field will of course depend on the motion chosen for K'. According to the general theory of relativity, the general law of the gravitational field must be satisfied for all gravitational fields obtainable in this

1. This follows from a generalisation of the discussion in Section 20.

way. Even though by no means all gravitational fields can be produced in this way, yet we may entertain the hope that the general law of gravitation will be derivable from such gravitational fields of a special kind. This hope has been realised in the most beautiful manner. But between the clear vision of this goal and its actual realisation it was necessary to surmount a serious difficulty, and as this lies deep at the root of things, I dare not withhold it from the reader. We require to extend our ideas of the space-time continuum still farther.

BEHAVIOUR OF CLOCKS AND MEASURING-RODS ON A ROTATING BODY OF REFERENCE†

Hitherto I have purposely refrained from speaking about the physical interpretation of space- and time-data in the case of the general theory of relativity. As a consequence, I am guilty of a certain slovenliness of treatment, which, as we know from the special theory of relativity, is far from being unimportant and pardonable. It is now high time that we remedy this defect; but I would mention at the outset, that this matter lays no small claims on the patience and on the power of abstraction of the reader.

We start off again from quite special cases, which we have frequently used before. Let us consider a space-time domain in which no gravitational fields exists relative to a reference-body K whose state of motion has been suitably chosen. K is then a Galileian reference-body as regards the domain considered, and the results of the special theory of relativity hold relative to K. Let us suppose the same domain referred to a second body of reference K', which is rotating uniformly with respect to K. In order to fix our ideas, we shall imagine K' to be in the form of a plane circular disc, which rotates uniformly in its own plane about its centre. An observer who is sitting eccentrically on the disc K' is sensible of a force which acts outwards in a radial direction, and which would be interpreted as an effect of inertia (centrifugal force) by an observer who was at rest with respect to the original reference-body K. But the observer on the disc may regard his disc as a reference-body which is "at rest"; on the basis of the general principle of relativity he is justified in doing this. The force acting on himself, and in fact on all other bodies which are at rest relative to the disc, he regards as the effect of a gravitational field. Nevertheless, the space-distribution of this gravitational field is of a kind that would not be possible on Newton's theory of gravitation.[1] But since the observer believes in the general theory of relativity, this does not disturb him; he is quite in the right when he believes that a general law of gravitation can be formulated—a law which not only explains the motion of the stars correctly, but also the field of force experienced by himself.

1. The field disappears at the centre of the disc and increases proportionally to the distance from the centre as we proceed outwards.

The observer performs experiments on his circular disc with clocks and measuring-rods. In doing so, it is his intention to arrive at exact definitions for the signification of time- and space-data with reference to the circular disc K', these definitions being based on his observations. What will be his experience in this enterprise?

To start with, he places one of two identically constructed clocks at the centre of the circular disc, and the other on the edge of the disc, so that they are at rest relative to it. We now ask ourselves whether both clocks go at the same rate from the standpoint of the non-rotating Galileian reference-body K. As judged from this body, the clock at the centre of the disc has no velocity, whereas the clock at the edge of the disc is in motion relative to K in consequence of the rotation. According to a result obtained in Section 12, it follows that the latter clock goes at a rate permanently slower than that of the clock at the centre of the circular disc, *i.e.* as observed from K. It is obvious that the same effect would be noted by an observer whom we will imagine sitting alongside his clock at the centre of the circular disc. Thus on our circular disc, or, to make the case more general, in every gravitational field, a clock will go more quickly or less quickly, according to the position in which the clock is situated (at rest). For this reason it is not possible to obtain a reasonable definition of time with the aid of clocks which are arranged at rest with respect to the body of reference. A similar difficulty presents itself when we attempt to apply our earlier definition of simultaneity in such a case, but I do not wish to go any farther into this question.

Moreover, at this stage the definition of the space co-ordinates also presents unsurmountable difficulties. If the observer applies his standard measuring-rod (a rod which is short as compared with the radius of the disc) tangentially to the edge of the disc, then, as judged from the Galileian system, the length of this rod will be less than 1, since, according to Section 12, moving bodies suffer a shortening in the direction of the motion. On the other hand, the measuring-rod will not experience a shortening in length, as judged from K, if it is applied to the disc in the direction of the radius. If, then, the observer first measures the circumference of the disc with his measuring-rod and then the diameter of the disc, on dividing the one by the other, he will not obtain as quotient the familiar number $\pi = 3.14$ ..., but a larger number,[2] whereas of course, for a disc which is at rest

2. Throughout this consideration we have to use the Galileian (non-rotating) system K as reference-body, since we may only assume the validity of the results of the special theory of relativity relative to K (relative to K' a gravitational field prevails).

with respect to K, this operation would yield π exactly. This proves that the propositions of Euclidean geometry cannot hold exactly on the rotating disc, nor in general in a gravitational field, at least if we attribute the length 1 to the rod in all positions and in every orientation. Hence the idea of a straight line also loses its meaning. We are therefore not in a position to define exactly the co-ordinates x, y, z relative to the disc by means of the method used in discussing the special theory, and as long as the co-ordinates and times of events have not been defined we cannot assign an exact meaning to the natural laws in which these occur.

Thus all our previous conclusions based on general relativity would appear to be called in question. In reality we must make a subtle detour in order to be able to apply the postulate of general relativity exactly. I shall prepare the reader for this in the following paragraphs.

EUCLIDEAN AND NON-EUCLIDEAN CONTINUUM

The surface of a marble table is spread out in front of me. I can get from any one point on this table to any other point by passing continuously from one point to a " neighbouring" one, and repeating this process a (large) number of times, or, in other words, by going from point to point without executing "jumps." I am sure the reader will appreciate with sufficient clearness what I mean here by "neighbouring" and by "jumps" (if he is not too pedantic). We express this property of the surface by describing the latter as a continuum.

Let us now imagine that a large number of little rods of equal length have been made, their lengths being small compared with the dimensions of the marble slab. When I say they are of equal length, I mean that one can be laid on any other without the ends overlapping. We next lay four of these little rods on the marble slab so that they constitute a quadrilateral figure (a square), the diagonals of which are equally long. To ensure the equality of the diagonals, we make use of a little testing-rod.

To this square we add similar ones, each of which has one rod in common with the first. We proceed in like manner with each of these squares until finally the whole marble slab is laid out with squares. The arrangement is such, that each side of a square belongs to two squares and each corner to four squares.

It is a veritable wonder that we can carry our this business without getting into the greatest difficulties. We only need to think of the following. If at any moment three squares meet at a corner, then two sides of the fourth square are already laid, and as a consequence, the arrangement of the remaining two sides of the square is already completely determined. But I am now no longer able to adjust the quadrilateral so that its diagonals may be equal. If they are equal of their own accord, then this is an especial favour of the marble slab and of the little rods about which I can only be thankfully surprised. We must needs experience many such surprises if the construction is to be successful.

If everything has really gone smoothly, then I say that the points of the marble slab constitute a Euclidean continuum with respect to the little rod, which has been used as a "distance" (line-interval). By choosing one corner of a square as "origin," I can characterise every other corner of a square with reference to this origin by means of two numbers. I only need state how many rods I must pass over when, starting from the origin, I proceed towards the "right" and then "upwards," in order to arrive at the corner of the square under consideration. These two numbers are then the "Cartesian co-ordinates" of this corner with reference to the "Cartesian co-ordinate system" which is determined by the arrangement of little rods.

By making use of the following modification of this abstract experiment, we recognise that there must also be cases in which the experiment would be unsuccessful. We shall suppose that the rods "expand" by an amount proportional to the increase of temperature. We heat the central part of the marble slab, but not the periphery, in which case two of our little rods can still be brought into coincidence at every position on the table. But our construction of squares must necessarily come into disorder during the heating, because the little rods on the central region of the table expand, whereas those on the outer part do not.

With reference to our little rods—defined as unit lengths—the marble slab is no longer a Euclidean continuum, and we are also no longer in the position of defining Cartesian co-ordinates directly with their aid, since the above construction can no longer be carried out. But since there are other things which are not influenced in a similar manner to the little rods (or perhaps not at all) by the temperature of the table, it is possible quite naturally to maintain the point of view that the marble slab is a "Euclidean continuum." This can be done in a satisfactory manner by making a more subtle stipulation about the measurement or the comparison of lengths.

But if rods of every kind (*i.e.* of every material) were to behave *in the same way* as regards the influence of temperature when they are on the variably heated marble slab, and if we had no other means of detecting the effect of temperature than the geometrical behaviour of our rods in experiments analogous to the one described above, then our best plan would be to assign the distance *one* to two points on the slab, provided that the ends of one of our rods could be made to coincide with these two points; for how else should we define the distance without our proceeding being in the highest measure grossly arbitrary? The method of Cartesian co-ordinates must then be discarded, and replaced by another which does not assume the validity of Euclidean geometry for rigid bodies.[1] The

---

1. Mathematicians have been confronted with our problem in the following form. If we are given a surface (*e.g.* an ellipsoid) in Euclidean three-dimensional space, then there exists for this surface a two-dimensional geometry, just as much as for a plane surface. Gauss undertook the task of treating

reader will notice that the situation depicted here corresponds to the one brought about by the general postulate of relativity (Section 23).

---

this two-dimensional geometry from first principles, without making use of the fact that the surface belongs to a Euclidean continuum of three dimensions. If we imagine constructions to be made with rigid rods *in the surface* (similar to that above with the marble slab), we should find that different laws hold for these from those resulting on the basis of Euclidean plane geometry. The surface is not a Euclidean continuum with respect to the rods, and we cannot define Cartesian co-ordinates *in the surface*. Gauss indicated the principles according to which we can treat the geometrical relationships in the surface, and thus pointed out the way to the method of Riemann of treating multi-dimensional, non-Euclidean *continua*. Thus it is that mathematicians long ago solved the formal problems to which we are led by the general postulate of relativity.

GAUSSIAN CO-ORDINATES

According to Gauss, this combined analytical and geometrical mode of handling the problem can be arrived at in the following way. We imagine a system of arbitrary curves (see Fig. 4) drawn on the surface of the table. These we designate as *u*-curves, and we indicate each of them by means of a number. The curves $u = 1$, $u = 2$ and $u = 3$ are drawn in the diagram. Between the curves $u = 1$ and $u = 2$ we must imagine an infinitely large number to be drawn, all of which correspond to real numbers lying between 1 and 2. We have then a system of

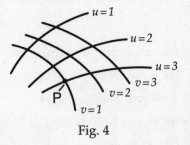

Fig. 4

u-curves, and this "infinitely dense" system covers the whole surface of the table. These *u*-curves must not intersect each other, and through each point of the surface one and only one curve must pass. Thus a perfectly definite value of *u* belongs to every point on the surface of the marble slab. In like manner we imagine a system of *v*-curves drawn on the surface. These satisfy the same conditions as the *u*-curves, they are provided with numbers in a corresponding manner, and they may likewise be of arbitrary shape. It follows that a value of *u* and a value of *v* belong to every point on the surface of the table. We call these two numbers the co-ordinates of the surface of the table (Gaussian co-ordinates). For example, the point *P* in the diagram has the Gaussian co-ordinates u=3, v=1. Two neighbouring points *P* and *P'* on the surface then correspond to the co-ordinates

$$P: \qquad u, v$$
$$P': \qquad u + du, v + dv,$$

where *du* and *dv* signify very small numbers. In a similar manner we may indicate the distance (line-interval) between *P* and *P'*, as measured with a little rod, by means of the very small number *ds*. Then according to Gauss we have

$$ds^2 = g_{11}du^2 + 2g_{12}dudv + g_{22}dv^2,$$

where g_{11}, g_{12}, g_{22}, are magnitudes which depend in a perfectly definite way on *u* and *v*. The magnitudes g_{11}, g_{12} and g_{22} determine the behaviour of the rods relative to the *u*-curves and *v*-curves, and thus also relative to the surface of the table. For the case in which the points of the surface considered form a Euclidean continuum with reference to the measuring-rods, but only in this case, it is possible to draw the u-curves and v-curves and to attach numbers to them, in such a manner, that we simply have:

$$ds^2 = du^2 + dv^2.$$

Under these conditions, the u-curves and v-curves are straight lines in the sense of Euclidean geometry, and they are perpendicular to each other. Here the Gaussian co-ordinates are simply Cartesian ones. It is clear that Gauss co-ordinates are nothing more than an association of two sets of numbers with the points of the surface considered, of such a nature that numerical values differing very slightly from each other are associated with neighbouring points "in space."

So far, these considerations hold for a continuum of two dimensions. But the Gaussian method can be applied also to a continuum of three, four or more dimensions. If, for instance, a continuum of four dimensions be supposed available, we may represent it in the following way. With every point of the continuum we associate arbitrarily four numbers, x_1, x_2, x_3, x_4, which are known as "co-ordinates." Adjacent points correspond to adjacent values of the co-ordinates. If a distance *ds* is associated with the adjacent points *P* and *P'*, this distance being measurable and well-defined from a physical point of view, then the following formula holds:

$$ds^2 = g_{11}dx_1{}^2 + 2g_{12}dx_1dx_2 \ldots + g_{44}dx_4{}^2,$$

where the magnitudes g_{11}, etc., have values which vary with the position in the continuum. Only when the continuum is a Euclidean one is it possible to associate the co-ordinates $x_1 \ldots x_4$ with the points of the continuum so that we have simply

$$ds^2 = dx_1{}^2 + dx_2{}^2 + dx_3{}^2 + dx_4{}^2$$

In this case relations hold in the four-dimensional continuum which are analogous to those holding in our three-dimensional measurements.

However, the Gauss treatment for ds^2 which we have given above is not always possible. It is only possible when sufficiently small regions of the continuum under consideration may be regarded as Euclidean continua. For example, this obviously holds in the case of the marble slab of the table and local variation of temperature. The temperature is practically constant for a small part of the slab, and thus the geometrical behaviour of the rods is *almost* as it ought to be according to the rules of

Euclidean geometry. Hence the imperfections of the construction of squares in the previous section do not show themselves clearly until this construction is extended over a considerable portion of the surface of the table.

We can sum this up as follows: Gauss invented a method for the mathematical treatment of continua in general, in which "size-relations" ("distances" between neighbouring points) are defined. To every point of a continuum are assigned as many numbers (Gaussian co-ordinates) as the continuum has dimensions. This is done in such a way, that only one meaning can be attached to the assignment, and that numbers (Gaussian co-ordinates) which differ by an indefinitely small amount are assigned to adjacent points. The Gaussian co-ordinate system is a logical generalisation of the Cartesian co-ordinate system. It is also applicable to non-Euclidean continua, but only when, with respect to the defined "size" or "distance," small parts of the continuum under consideration behave more nearly like a Euclidean system, the smaller the part of the continuum under our notice.

26

THE SPACE-TIME CONTINUUM OF THE SPECIAL THEORY OF RELATIVITY CONSIDERED AS A EUCLIDEAN CONTINUUM

We are now in a position to formulate more exactly the idea of Minkowski, which was only vaguely indicated in Section 17. In accordance with the special theory of relativity, certain co-ordinate systems are given preference for the description of the four-dimensional, space-time continuum. We called these "Galileian co-ordinate systems." For these systems, the four co-ordinates x, y, z, t, which determine an event or—in other words—a point of the four-dimensional continuum, are defined physically in a simple manner, as set forth in detail in the first part of this book. For the transition from one Galileian system to another, which is moving uniformly with reference to the first, the equations of the Lorentz transformation are valid. These last form the basis for the derivation of deductions from the special theory of relativity, and in themselves they are nothing more than the expression of the universal validity of the law of transmission of light for all Galileian systems of reference.

Minkowski found that the Lorentz transformations satisfy the following simple conditions. Let us consider two neighbouring events, the relative position of which in the four-dimensional continuum is given with respect to a Galileian reference-body K by the space co-ordinate differences dx, dy, dz and the time-difference dt. With reference to a second Galileian system we shall suppose that the corresponding differences for these two events are dx', dy', dz', dt'. Then these magnitudes always fulfil the condition.[1]

$$dx^2 + dy^2 + dz^2 - c^2\, dt^2 = dx'^2 + dy'^2 + dz'^2 - c'^2\, dt'^2.$$

The validity of the Lorentz transformation follows from this condition. We can express this as follows: The magnitude

$$ds^2 = dx^2 + dy^2 + dz^2 - c^2\, dt^2,$$

which belongs to two adjacent points of the four-dimensional space-time continuum, has the same value for all selected (Galileian) reference-bodies. If we

1. Cf. Appendices 1 and 2. The relations which are derived there for the co-ordinates themselves are valid also for co-ordinate *differences*, and thus also for co-ordinate differentials (indefinitely small differences).

replace *x, y, z,* $\sqrt{-1}$ *ct,* by x_1, x_2, x_3, x_4, we also obtain the result that

$$ds^2 = dx_1{}^2 + dx_2{}^2 + dx_3{}^3 + dx_4{}^2$$

is independent of the choice of the body of reference. We call the magnitude *ds* the "distance" apart of the two events or four-dimensional points.

Thus, if we choose as time-variable the imaginary variable $\sqrt{-1}$ *ct* instead of the real quantity *t,* we can regard the space-time continuum—in accordance with the special theory of relativity—as a "Euclidean" four-dimensional continuum, a result which follows from the considerations of the preceding section.

27

THE SPACE-TIME CONTINUUM OF THE GENERAL THEORY OF RELATIVITY IS NOT A EUCLIDEAN CONTINUUM†

In the first part of this book we were able to make use of space-time co-ordinates which allowed of a simple and direct physical interpretation, and which, according to Section 26, can be regarded as four-dimensional Cartesian co-ordinates. This was possible on the basis of the law of the constancy of the velocity of light. But according to Section 21, the general theory of relativity cannot retain this law. On the contrary, we arrived at the result that according to this latter theory the velocity of light must always depend on the coordinates when a gravitational field is present. In connection with a specific illustration in Section 23, we found that the presence of a gravitational field invalidates the definition of the co-ordinates and the time, which led us to our objective in the special theory of relativity.

In view of the results of these considerations we are led to the conviction that, according to the general principle of relativity, the space-time continuum cannot be regarded as a Euclidean one, but that here we have the general case, corresponding to the marble slab with local variations of temperature, and with which we made acquaintance as an example of a two-dimensional continuum. Just as it was there impossible to construct a Cartesian co-ordinate system from equal rods, so here it is impossible to build up a system (reference-body) from rigid bodies and clocks, which shall be of such a nature that measuring-rods and clocks, arranged rigidly with respect to one another, shall indicate position and time directly. Such was the essence of the difficulty with which we were confronted in Section 23.

But the considerations of Sections 25 and 26 show us the way to surmount this difficulty. We refer the four-dimensional space-time continuum in an arbitrary manner to Gauss co-ordinates. We assign to every point of the continuum (event) four numbers, x_1, x_2, x_3, x_4 (co-ordinates), which have not the least direct physical significance, but only serve the purpose of numbering the points of the continuum in a definite but arbitrary manner. This arrangement does not even need to be of such a kind that we must regard x_1, x_2, x_3 as "space" co-ordinates and x_4 as a "time" co-ordinate.

The reader may think that such a description of the world would be quite inadequate. What does it mean to assign to an event the particular co-ordinates x_1, x_2, x_3, x_4, if in themselves these co-ordinates have no significance? More careful consideration shows, however, that this anxiety is unfounded. Let us consider, for instance, a material point with any kind of motion. If this point had only a momentary existence without duration, then it would be described in space-time by a single system of values x_1, x_2, x_3, x_4. Thus its permanent existence must be characterised by an infinitely large number of such systems of values, the co-ordinate values of which are so close together as to give continuity; corresponding to the material point, we thus have a (uni-dimensional) line in the four-dimensional continuum. In the same way, any such lines in our continuum correspond to many points in motion. The only statements having regard to these points which can claim a physical existence are in reality the statements about their encounters. In our mathematical treatment, such an encounter is expressed in the fact that the two lines which represent the motions of the points in question have a particular system of co-ordinate values, x_1, x_2, x_3, x_4, in common. After mature consideration the reader will doubtless admit that in reality such encounters constitute the only actual evidence of a time-space nature with which we meet in physical statements.

When we were describing the motion of a material point relative to a body of reference, we stated nothing more than the encounters of this point with particular points of the reference-body. We can also determine the corresponding values of the time by the observation of encounters of the body with clocks, in conjunction with the observation of the encounter of the hands of clocks with particular points on the dials. It is just the same in the case of space-measurements by means of measuring-rods, as a little consideration will show.

The following statements hold generally: Every physical description resolves itself into a number of statements, each of which refers to the space-time coincidence of two events A and B. In terms of Gaussian co-ordinates, every such statement is expressed by the agreement of their four co-ordinates x_1, x_2, x_3, x_4. Thus in reality, the description of the time-space continuum by means of Gauss co-ordinates completely replaces the description with the aid of a body of reference, without suffering from the defects of the latter mode of description; it is not tied down to the Euclidean character of the continuum which has to be represented.

EXACT FORMULATION OF THE GENERAL PRINCIPLE OF RELATIVITY†

W e are now in a position to replace the provisional formulation of the general principle of relativity given in Section 28 by an exact formulation. The form there used, "All bodies of reference K, K', etc., are equivalent for the description of natural phenomena (formulation of the general laws of nature), whatever may be their state of motion," cannot be maintained, because the use of rigid reference-bodies, in the sense of the method followed in the special theory of relativity, is in general not possible in space-time description. The Gauss co-ordinate system has to take the place of the body of reference. The following statement corresponds to the fundamental idea of the general principle of relativity: *"All Gaussian co-ordinate systems are essentially equivalent for the formulation of the general laws of nature."*

We can state this general principle of relativity in still another form, which renders it yet more clearly intelligible than it is when in the form of the natural extension of the special principle of relativity. According to the special theory of relativity, the equations which express the general laws of nature pass over into equations of the same form when, by making use of the Lorentz transformation, we replace the space-time variables x, y, z, t, of a (Galileian) reference-body K by the space-time variables x', y', z', t', of a new reference-body K'. According to the general theory of relativity, on the other hand, by application of *arbitrary substitutions* of the Gauss variables x_1, x_2, x_3, x_4, the equations must pass over into equations of the same form; for every transformation (not only the Lorentz transformation) corresponds to the transition of one Gauss co-ordinate system into another.

If we desire to adhere to our "old-time" three-dimensional view of things, then we can characterise the development which is being undergone by the fundamental idea of the general theory of relativity as follows: The special theory of relativity has reference to Galileian domains, *i.e.* to those in which no gravitational field exists. In this connection a Galileian reference-body serves as body of reference, *i.e.* a rigid body the state of motion of which is so chosen that the Galileian law of the uniform rectilinear motion of "isolated" material points holds relatively to it.

Certain considerations suggest that we should refer the same Galileian domains to *non-Galileian* reference-bodies also. A gravitational field of a special kind is then present with respect to these bodies (cf. Sections 20 and 23).

In gravitational fields there are no such things as rigid bodies with Euclidean properties; thus the fictitious rigid body of reference is of no avail in the general theory of relativity. The motion of clocks is also influenced by gravitational fields, and in such a way that a physical definition of time which is made directly with the aid of clocks has by no means the same degree of plausibility as in the special theory of relativity.

For this reason non-rigid reference-bodies are used which are as a whole not only moving in any way whatsoever, but which also suffer alterations in form *ad lib.* during their motion. Clocks, for which the law of motion is any kind, however irregular, serve for the definition of time. We have to imagine each of these clocks fixed at a point on the non-rigid reference-body. These clocks satisfy only the one condition, that the "readings" which are observed simultaneously on adjacent clocks (in space) differ from each other by an indefinitely small amount. This non-rigid reference-body, which might appropriately be termed a "reference-mollusk," is in the main equivalent to a Gaussian four-dimensional co-ordinate system chosen arbitrarily. That which gives the "mollusk" a certain comprehensibleness as compared with the Gauss co-ordinate system is the (really unqualified) formal retention of the separate existence of the space co-ordinate. Every point on the mollusk is treated as a space-point, and every material point which is at rest relatively to it as at rest, so long as the mollusk is considered as reference-body. The general principle of relativity requires that all these mollusks can be used as reference-bodies with equal right and equal success in the formulation of the general laws of nature; the laws themselves must be quite independent of the choice of mollusk.

The great power possessed by the general principle of relativity lies in the comprehensive limitation which is imposed on the laws of nature in consequence of what we have seen above.

THE SOLUTION OF THE PROBLEM OF GRAVITATION ON THE BASIS OF THE GENERAL PRINCIPLE OF RELATIVITY

If the reader has followed all our previous considerations, he will have no further difficulty in understanding the methods leading to the solution of the problem of gravitation.

We start off from a consideration of a Galileian domain, *i.e.* a domain in which there is no gravitational field relative to the Galileian reference-body K. The behaviour of measuring-rods and clocks with reference to K is known from the special theory of relativity, likewise the behaviour of "isolated" material points; the latter move uniformly and in straight lines.

Now let us refer this domain to a random Gauss coordinate system or to a "mollusk" as reference-body K'. Then with respect to K' there is a gravitational field G (of a particular kind). We learn the behaviour of measuring-rods and clocks and also of freely-moving material points with reference to K' simply by mathematical transformation. We interpret this behaviour as the behaviour of measuring-rods, clocks and material points under the influence of the gravitational field *G*. Hereupon we introduce a hypothesis: that the influence of the gravitational field on measuring-rods, clocks and freely-moving material points continues to take place according to the same laws, even in the case when the prevailing gravitational field is *not* derivable from the Galileian special case, simply by means of a transformation of co-ordinates.

The next step is to investigate the space-time behaviour of the gravitational field *G*, which was derived from the Galileian special case simply by transformation of the co-ordinates. This behaviour is formulated in a law, which is always valid, no matter how the reference-body (mollusk) used in the description may be chosen.

This law is not yet the *general* law of the gravitational field, since the gravitational field under consideration is of a special kind. In order to find out the general law-of-field of gravitation we still require to obtain a generalisation of the law as found above. This can be obtained without caprice, however, by taking into consideration the following demands:

> (*a*) The required generalisation must likewise satisfy the general postulate of relativity.

(*b*) If there is any matter in the domain under consideration, only its inertial mass, and thus according to Section 15 only its energy is of importance for its effect in exciting a field.

(*c*) Gravitational field and matter together must satisfy the law of the conservation of energy (and of impulse).

Finally, the general principle of relativity permits us to determine the influence of the gravitational field on the course of all those processes which take place according to known laws when a gravitational field is absent, *i.e.* which have already been fitted into the frame of the special theory of relativity. In this connection we proceed in principle according to the method which has already been explained for measuring-rods, clocks and freely-moving material points.

The theory of gravitation derived in this way from the general postulate of relativity excels not only in its beauty; nor in removing the defect attaching to classical mechanics which was brought to light in Section 21; nor in interpreting the empirical law of the equality of inertial and gravitational mass; but it has also already explained a result of observation in astronomy, against which classical mechanics is powerless.

If we confine the application of the theory to the case where the gravitational fields can be regarded as being weak, and in which all masses move with respect to the co-ordinate system with velocities which are small compared with the velocity of light, we then obtain as a first approximation the Newtonian theory. Thus the latter theory is obtained here without any particular assumption, whereas Newton had to introduce the hypothesis that the force of attraction between mutually attracting material points is inversely proportional to the square of the distance between them. If we increase the accuracy of the calculation, deviations from the theory of Newton make their appearance, practically all of which must nevertheless escape the test of observation owing to their smallness.

We must draw attention here to one of these deviations. According to Newton's theory, a planet moves round the sun in an ellipse, which would permanently maintain its position with respect to the fixed stars, if we could disregard the motion of the fixed stars themselves and the action of the other planets under consideration. Thus, if we correct the observed motion of the planets for these two influences, and if Newton's theory be strictly correct, we ought to obtain for the orbit of the planet an ellipse, which is fixed with reference to the fixed stars. This deduction, which can be tested with great accuracy, has been confirmed for all the planets save one, with the precision that is capable of being obtained by the delicacy of observation attainable at the present time. The sole exception is Mercury, the planet which lies nearest the sun. Since the time Leverrier, it has been known that the ellipse corresponding to the orbit of Mercury, after it has been corrected for the influences mentioned above, is not stationary with respect to the fixed stars, but that it rotates exceedingly slowly in the plane of the orbit and in the sense of the orbital motion. The value obtained for this rotary movement of the orbital ellipse was 43 seconds of arc per century, an amount ensured to be correct to within a few seconds of arc. This effect can be explained by means of classical mechanics only on the assumption of hypotheses which have little probability, and which were devised solely for this purpose.

On the basis of the general theory of relativity, it is found that the ellipse of every planet round the sun must necessarily rotate in the manner indicated above; that for all the planets, with the exception of Mercury, this rotation is too small to be detected with the delicacy of observation possible at the present time; but that in the case of Mercury it must amount to 43 seconds of arc per century, a result which is strictly in agreement with observation.

Apart from this one, it has hitherto been possible to make only two deductions from the theory which admit of being tested by observation, to wit, the curvature of light rays by the gravitational field of the sun,[1] and a displacement of the spectral lines of light reaching us from large stars, as compared with the corresponding lines for light produced in an analogous manner terrestrially (*i.e.* by the same kind of molecule). I do not doubt that these deductions from the theory will be confirmed also.

1. Observed by Eddington and others in 1919. (Cf. Appendix 3.)

PART III

CONSIDERATIONS ON THE UNIVERSE AS A WHOLE

30

COSMOLOGICAL DIFFICULTIES OF NEWTON'S THEORY

Apart from the difficulty discussed in Section 21, there is a second fundamental difficulty attending classical celestial mechanics, which, to the best of my knowledge, was first discussed in detail by the astronomer Seeliger. If we ponder over the question as to how the universe, considered as a whole, is to be regarded, the first answer that suggests itself to us is surely this: As regards space (and time) the universe is infinite. There are stars everywhere, so that the density of matter, although very variable in detail, is nevertheless on the average everywhere the same. In other words: However far we might travel through space, we should find everywhere an attenuated swarm of fixed stars of approximately the same kind and density.

This view is not in harmony with the theory of Newton. The latter theory rather requires that the universe should have a kind of centre in which the density of the stars is a maximum, and that as we proceed outwards from this centre the group-density of the stars should diminish, until finally, at great distances, it is succeeded by an infinite region of emptiness. The stellar universe ought to be a finite island in the infinite ocean of space.[1]

This conception is in itself not very satisfactory. It is still less satisfactory because it leads to the result that the light emitted by the stars and also individual stars of the stellar system are perpetually passing out into infinite space, never to return, and without ever again coming into interaction with other objects of nature. Such a

1. *Proof*—According to the theory of Newton, the number of "lines of force" which come from infinity and terminate in a mass m is proportional to the mass m. If, on the average, the mass-density ρ_0 is constant throughout the universe, then a sphere of volume V will enclose the average mass $\rho_0 V$. Thus the number of lines of force passing through the surface F of the sphere into its interior is proportional to $\rho_0 V$. For unit area of the surface of the sphere the number of lines of force which enters the sphere is thus proportional to $\rho_0 V/F$ or $\rho_0 R$. Hence the intensity of the field at the surface would ultimately become infinite with increasing radius R of the sphere, which is impossible.

finite material universe would be destined to become gradually but systematically impoverished.

In order to escape this dilemma, Seeliger suggested a modification of Newton's law, in which he assumes that for great distances the force of attraction between two masses diminishes more rapidly than would result from the inverse square law. In this way it is possible for the mean density of matter to be constant everywhere, even to infinity, without infinitely large gravitational fields being produced. We thus free ourselves from the distasteful conception that the material universe ought to possess something of the nature of centre. Of course we purchase our emancipation from the fundamental difficulties mentioned, at the cost of a modification and complication of Newton's law which has neither empirical nor theoretical foundation. We can imagine innumerable laws which would serve the same purpose, without our being able to state a reason why one of them is to be preferred to the others; for any one of these laws would be founded just as little on more general theoretical principles as is the law of Newton.

31

THE POSSIBILITY OF A "FINITE" AND YET "UNBOUNDED" UNIVERSE

But speculations on the structure of the universe also move in quite another direction. The development of non-Euclidean geometry led to the recognition of the fact, that we can cast doubt on the *infiniteness* of our space without coming into conflict with the laws of thought or with experience (Riemann, Helmholtz). These questions have already been treated in detail and with unsurpassable lucidity by Helmholtz and Poincaré, whereas I can only touch on them briefly here.

In the first place, we imagine an existence in two-dimensional space. Flat beings with flat implements, and in particular flat rigid measuring-rods, are free to move in a *plane*. For them nothing exists outside of this plane: that which they observe to happen to themselves and to their flat "things" is the all-inclusive reality of their plane. In particular, the constructions of plane Euclidean geometry can be carried out by means of the rods, *e.g.* the lattice construction, considered in Section

24. In contrast to ours, the universe of these beings is two-dimensional; but, like ours, it extends to infinity. In their universe there is room for an infinite number of identical squares made up of rods, *i.e.* its volume (surface) is infinite. If these beings say their universe is "plane," there is sense in the statement, because they mean that they can perform the constructions of plane Euclidean geometry with their rods. In this connection the individual rods always represent the same distance, independently of their position.

Let us consider now a second two-dimensional existence, but this time on a spherical surface instead of on a plane. The flat beings with their measuring-rods and other objects fit exactly on this surface and they are unable to leave it. Their whole universe of observation extends exclusively over the surface of the sphere. Are these beings able to regard the geometry of their universe as being plane geometry and their rods withal as the realisation of "distance"? They cannot do this. For if they attempt to realise a straight line, they will obtain a curve, which we "three-dimensional beings" designate as a great circle, *i.e.* a self-contained line of definite finite length, which can be measured up by means of a measuring-rod. Similarly, this universe has a finite area, that can be compared with the area of a square constructed with rods. The great charm resulting from this consideration lies in the recognition of the fact that *the universe of these beings is finite and yet has no limits.*

But the spherical-surface beings do not need to go on a world-tour in order to perceive that they are not living in a Euclidean universe. They can convince themselves of this on every part of their "world," provided they do not use too small a piece of it. Starting from a point, they draw "straight lines" (arcs of circles as judged in three-dimensional space) of equal length in all directions. They will call the line joining the free ends of these lines a "circle." For a plane surface, the ratio of the circumference of a circle to its diameter, both lengths being measured with the same rod, is, according to Euclidean geometry of the plane, equal to a constant value π, which is independent of the diameter of the circle. On their spherical surface our flat beings would find for this ratio the value

$$\pi \frac{\sin\left(\dfrac{r}{R}\right)}{\left(\dfrac{r}{R}\right)},$$

i.e. a smaller value than π, the difference being the more considerable, the greater is the radius of the circle in comparison with the radius R of the "world-sphere." By means of this relation the spherical beings can determine the radius of their universe ("world"), even when only a relatively small part of their world-sphere is available for their measurements. But if this part is very small indeed, they will no longer be able to demonstrate that they are on a spherical "world" and not on a Euclidean plane, for a small part of a spherical surface differs only slightly from a piece of a plane of the same size.

Thus if the spherical-surface beings are living on a planet of which the solar system occupies only a negligibly small part of the spherical universe, they have no means of determining whether they are living in a finite or in an infinite universe, because the "piece of universe" to which they have access is in both cases practically plane, or Euclidean. It follows directly from this discussion, that for our sphere-beings the circumference of a circle first increases with the radius until the "circumference of the universe" is reached, and that it thenceforward gradually decreases to zero for still further increasing values of the radius. During this process the area of the circle continues to increase more and more, until finally it becomes equal to the total area of the whole "world-sphere."

Perhaps the reader will wonder why we have placed our "beings" on a sphere rather than on another closed surface. But this choice has its justification in the fact that, of all closed surfaces, the sphere is unique in possessing the property that all points on it are equivalent. I admit that the ratio of the circumference c of a circle to its radius r depends on r, but for a given value of r it is the same for all points of the "world-sphere"; in other words, the "world-sphere" is a "surface of constant curvature."

To this two-dimensional sphere-universe there is a three-dimensional analogy, namely, the three-dimensional spherical space which was discovered by Riemann. Its points are likewise all equivalent. It possesses a finite volume, which is determined by its "radius" $(2\pi^2 R^3)$. Is it possible to imagine a spherical space? To imagine a space means nothing else than that we imagine an epitome of our "space" experience, $i.e.$ of experience that we can have in the movement of "rigid" bodies. In this sense we *can* imagine a spherical space.

Suppose we draw lines or stretch strings in all directions from a point, and mark off from each of these the distance r with a measuring-rod. All the free end-points of these lengths lie on a spherical surface. We can specially measure up the area (F) of this surface by means of a square made up of measuring-rods. If the universe is Euclidean, then $F = 4\pi r^2$; if it is spherical, then F is always less than $4\pi r^2$. With increasing values of r, F increases from zero up to a maximum value which is determined by the "world-radius," but for still further increasing values of r, the area gradually diminishes to zero. At first, the straight lines which radiate from the starting point diverge farther and farther from one another, but later they approach each other, and finally they run together again at a "counter-point" to the starting point. Under such conditions they have traversed the whole spherical space. It is easily seen that the three-dimensional spherical space is quite analogous to the two-dimensional spherical surface. It is finite (*i.e.* of finite volume), and has no bounds.

It may be mentioned that there is yet another kind of curved space: "elliptical space." It can be regarded as a curved space in which the two "counter-points" are identical (indistinguishable from each other). An elliptical universe can thus be considered to some extent as a curved universe possessing central symmetry.

It follows from what has been said, that closed spaces without limits are conceivable. From amongst these, the spherical space (and the elliptical) excels in its simplicity, since all points on it are equivalent. As a result of this discussion, a most interesting question arises for astronomers and physicists, and that is whether the universe in which we live is infinite, or whether it is finite in the manner of the spherical universe. Our experience is far from being sufficient to enable us to answer this question. But the general theory of relativity permits of our answering it with a moderate degree of certainty, and in this connection the difficulty mentioned in Section 30 finds its solution.

THE STRUCTURE OF SPACE ACCORDING TO THE GENERAL THEORY OF RELATIVITY†

According to the general theory of relativity, the geometrical properties of space are not independent, but they are determined by matter. Thus we can draw conclusions about the geometrical structure of the universe only if we base our considerations on the state of the matter as being something that is known. We know from experience that, for a suitably chosen co-ordinate system, the velocities of the stars are small as compared with the velocity of transmission of light. We can thus as a rough approximation arrive at a conclusion as to the nature of the universe as a whole, if we treat the matter as being at rest.

We already know from our previous discussion that the behaviour of measuring-rods and clocks is influenced by gravitational fields, *i.e.* by the distribution of matter. This in itself is sufficient to exclude the possibility of the exact validity of Euclidean geometry in our universe. But it is conceivable that our universe differs only slightly from a Euclidean one, and this notion seems all the more probable, since calculations show that the metrics of surrounding space is influenced only to an exceedingly small extent by masses even of the magnitude of our sun. We might imagine that, as regards geometry, our universe behaves analogously to a surface which is irregularly curved in its individual parts, but which nowhere departs appreciably from a plane: something like the rippled surface of a lake. Such a universe might fittingly be called a quasi-Euclidean universe. As regards its space it would be infinite. But calculation shows that in a quasi-Euclidean universe the average density of matter would necessarily be *nil*. Thus such a universe could not be inhabited by matter everywhere; it would present to us that unsatisfactory picture which we portrayed in Section 30.

If we are to have in the universe an average density of matter which differs from zero, however small may be that difference, then the universe cannot be quasi-Euclidean. On the contrary, the results of calculation indicate that if matter be distributed uniformly, the universe would necessarily be spherical (or elliptical). Since in reality the detailed distribution of matter is not uniform, the real universe will deviate in individual parts from the spherical, *i.e.* the universe will be quasi-spherical. But it will be necessarily finite. In fact, the theory supplies us with a simple connection[1] between the space-expanse of the universe and the average density of matter in it.

1. For the "radius" R of the universe we obtain the equation

$$R^2 = \frac{2}{k\rho}$$

The use of the C.G.S. system in this equation gives $2/k = 1.08 \cdot 10^{27}$; ρ is the average density of the matter.

SIMPLE DERIVATION OF THE LORENTZ TRANSFORMATION

[SUPPLEMENTARY TO SECTION 11]

For the relative orientation of the co-ordinate systems indicated in Fig. 2, the x-axes of both systems permanently coincide. In the present case we can divide the problem into parts by considering first only events which are localised on the x-axis. Any such event is represented with respect to the co-ordinate system K by the abscissa x and the time t, and with respect to the system K' by the abscissa x' and the time t'. We require to find x' and t' when x and t are given.

A light-signal, which is proceeding along the positive axis of x, is transmitted according to the equation

$$x = ct$$

or

$$x - ct = 0 \dots \dots (1).$$

Since the same light-signal has to be transmitted relative to K' with the velocity c, the propagation relative to the system K' will be represented by the analogous formula

$$x' - ct' = o \; . \; . \; . \; . \; . \; . \; . \; .(2).$$

Those space-time points (events) which satisfy (1) must also satisfy (2). Obviously this will be the case when the relation

$$(x' - ct') = \lambda(x - ct) \; . \; . \; . \; . \; . \; (3)$$

is fulfilled in general, where λ indicates a constant; for, according to (3), the disappearance of $(x - ct)$ involves the disappearance of $(x' - ct')$.

If we apply quite similar considerations to light rays which are being transmitted along the negative x-axis, we obtain the condition

$$(x' + ct') = \mu(x + ct) \; . \; . \; . \; . \; . \; (4).$$

By adding (or subtracting) equations (3) and (4), and introducing for convenience the constants a and b in place of the constants λ and μ where

$$a = \frac{\lambda + \mu}{2}$$

and

$$b = \frac{\lambda - \mu}{2},$$

we obtain the equations

$$x' = ax - bct \\ ct' = act - bx \biggr\} \cdots \cdots (5).$$

We should thus have the solution of our problem, if the constants a and b were known. These result from the following discussion.

For the origin of K' we have permanently $x' = 0$, and hence according to the first of the equations (5)

$$x = \frac{bc}{a}t.$$

If we call v the velocity with which the origin of K' is moving relative to K, we then have

$$v = \frac{bc}{a}. \cdots \cdots \cdots (6).$$

The same value v can be obtained from equation (5), if we calculate the velocity of another point of K' relative to K, or the velocity (directed towards the negative x-axis) of a point of K with respect to K'. In short, we can designate v as the relative velocity of the two systems.

Furthermore, the principle of relativity teaches us that, as judged from K, the length of a unit measuring-rod which is at rest with reference to K' must be exactly the same as the length, as judged from K', of a unit measuring-rod which is at rest relative to K. In order to see how the points of the x'-axis appear as viewed from K, we only require to take a "snapshot" of K' from K; this means that we have to insert a particular value of t (time of K), e.g. $t = 0$. For this value of t we then obtain from the first of the equations (5)

$$x' = ax.$$

Two points of the x'-axis which are separated by the distance $\Delta x' = 1$ when measured in the K' system are thus separated in our instantaneous photograph by the distance

$$\Delta x = \frac{1}{a}. \ldots \ldots \ldots (7).$$

But if the snapshot be taken from $K'(t' = 0)$, and if we eliminate t from the equations (5), taking into account the expression (6), we obtain

$$x' = a\left(1 - \frac{v^2}{c^2}\right)x.$$

From this we conclude that two points on the x-axis and separated by the distance 1 (relative to K) will be represented on our snapshot by the distance

$$\Delta x' = a\left(1 - \frac{v^2}{c^2}\right) \ldots \ldots (7a).$$

But from what has been said, the two snapshots must be identical; hence Δx in (7) must be equal to $\Delta x'$ in (7a), so that we obtain

$$a^2 = \frac{1}{1 - \dfrac{v^2}{c^2}} \ \dots\dots\dots (7b).$$

The equations (6) and (7b) determine the constants a and b. By inserting the values of these constants in (5), we obtain the first and the fourth of the equations given in Section 11.

$$\left. \begin{array}{l} x' = \dfrac{x - vt}{\sqrt{1 - \dfrac{v^2}{c^2}}} \\[3em] t' = \dfrac{t - \dfrac{v}{c^2}\,x}{\sqrt{1 - \dfrac{v^2}{c^2}}} \end{array} \right\} \dots\dots (8).$$

Thus we have obtained the Lorentz transformation for events on the x-axis. It satisfies the condition

$$x'^2 - c^2 t'^2 = x^2 - c^2 t^2 \dots\dots (8a).$$

The extension of this result, to include events which take place outside the x-axis, is obtained by retaining equations (8) and supplementing them by the relations

$$\left. \begin{array}{l} y' = y \\ z' = z \end{array} \right\} \dots\dots\dots (9).$$

In this way we satisfy the postulate of the constancy of the velocity of light *in vacuo* for rays of light of arbitrary direction, both for the system K and for the system K'. This may be shown in the following manner.

We suppose a light-signal sent out from the origin of K at the time $t = 0$. It will be propagated according to the equation

$$r = \sqrt{x^2 + y^2 + z^2} = ct,$$

or, if we square this equation, according to the equation

$$x^2 + y^2 + z^2 - c^2 t^2 = 0. \ldots (10).$$

It is required by the law of propagation of light, in conjunction with the postulate of relativity, that the transmission of the signal in question should take place—as judged from K'—in accordance with the corresponding formula

$$r' = ct'$$

or,

$$x'^2 + y'^2 + z'^2 - c^2 t'^2 = 0. \ldots (10a).$$

In order that equation (10a) may be a consequence of equation (10), we must have

$$x'^2 + y'^2 + z'^2 - c^2 t'^2 = \sigma(x^2 + y^2 + z^2 - c^2 t^2) \quad (11).$$

Since equation (8a) must hold for points on the x-axis, we thus have $\sigma = 1$. It is easily seen that the Lorentz transformation really satisfies equation (11) for $\sigma = 1$; for (11) is a consequence of (8a) and (9), and hence also of (8) and (9). We have thus derived the Lorentz transformation.

The Lorentz transformation represented by (8) and (9) still requires to be generalised. Obviously it is immaterial whether the axes of K' be chosen so that they are spatially parallel to those of K. It is also not essential that the velocity of translation of K' with respect to K should be in the direction of the x-axis. A simple consideration shows that we are able to construct the Lorentz transformation in this general sense from two kinds of transformations, viz. from Lorentz transformations in the special sense and from purely spatial transformations, which corresponds to the replacement of the rectangular co-ordinate system by a new system with its axes pointing in other directions.

Mathematically, we can characterise the generalised Lorentz transformation thus:

It expresses x', y', z', t', in terms of linear homogeneous functions of x, y, z, t, of such a kind that the relation

$$x'^2 + y'^2 + z'^2 - c^2 t'^2 = x^2 + y^2 + z^2 - c^2 t^2 \ . \ (11a)$$

is satisfied identically. That is to say: If we substitute their expressions in x, y, z, t, in place of x', y', z', t', on the left-hand side, then the left-hand side of (11a) agrees with the right-hand side.

MINKOWSKI'S FOUR-DIMENSIONAL SPACE ("WORLD")

[SUPPLEMENTARY TO SECTION 17]

We can characterise the Lorentz transformation still more simply if we introduce the imaginary $\sqrt{-1}$. ct in place of t, as time-variable. If, in accordance with this, we insert

$$x_1 = x$$
$$x_2 = y$$
$$x_3 = z$$
$$x_4 = \sqrt{-1}.ct,$$

and similarly for the accented system K', then the condition which is identically satisfied by the transformation can be expressed thus:

$$x_1'^2 + x_2'^2 + x_3'^2 + x_4'^2 = x_1^2 + x_2^2 + x_3^2 + x_4^2. \quad (12).$$

That is, by the afore-mentioned choice of "co-ordinates" ($11a$) is transformed into this equation.

We see from (12) that the imaginary time co-ordinate x_4 enters into the condition of transformation in exactly the same way as the space co-ordinates x_1, x_2, x_3. It is due to this fact that, according to the theory of relativity, the "time" x_4 enters into natural laws in the same form as the space co-ordinates x_1, x_2, x_3.

A four-dimensional continuum described by the "co-ordinates" x_1, x_2, x_3, x_4, was called "world" by Minkowski, who also termed a point-event a "world-point." From a "happening" in three-dimensional space, physics becomes, as it were, an "existence" in the four-dimensional "world."

This four-dimensional "world" bears a close similarity to the three-dimensional "space" of (Euclidean) analytical geometry. If we introduce into the latter a new Cartesian co-ordinate system (x'_1, x'_2, x'_3) with the same origin, then x'_1, x'_2, x'_3, are linear homogeneous functions of x_1, x_2, x_3, which identically satisfy the equation

$$x_1'^2 + x_2'^2 + x_3'^2 = x_1^2 + x_2^2 + x_3^2.$$

The analogy with (12) is a complete one. We can regard Minkowski's "world" in a formal manner as a four-dimensional Euclidean space (with imaginary time co-ordinate); the Lorentz transformation corresponds to a "rotation" of the co-ordinate system in the four-dimensional "world."

THE EXPERIMENTAL CONFIRMATION OF THE GENERAL THEORY OF RELATIVITY

From a systematic theoretical point of view, we may imagine the process of evolution of an empirical science to be a continuous process of induction. Theories are evolved, and are expressed in short compass as statements of a large number of individual observations in the form of empirical laws, from which the general laws can be ascertained by comparison. Regarded in this way, the development of a science bears some resemblance to the compilation of a classified catalogue. It is, as it were, a purely empirical enterprise.

But this point of view by no means embraces the whole of the actual process; for it slurs over the important part played by intuition and deductive thought in the development of an exact science. As soon as a science has emerged from its initial stages, theoretical advances are no longer achieved merely by a process of arrangement. Guided by empirical data, the investigator rather develops a system of thought which, in general, is built up logically from a small number of fundamental assumptions, the so-called axioms. We call such a system of thought a *theory*. The theory finds the justification for its existence in the fact that it correlates a large number of single observations, and it is just here that the "truth" of the theory lies.

Corresponding to the same complex of empirical data, there may be several theories, which differ from one another to a considerable extent. But as regards the deductions from the theories which are capable of being tested, the agreement between the theories may be so complete, that it becomes difficult to find such deductions in which the two theories differ from each other. As an example, a case of general interest is available in the province of biology, in the Darwinian theory of the development of species by selection in the struggle for existence, and in the theory of development which is based on the hypothesis of the hereditary transmission of acquired characters.

We have another instance of far-reaching agreement between the deductions from two theories in Newtonian mechanics on the one hand, and the general theory of relativity on the other. This agreement goes so far, that up to the present we have been able to find only a few deductions from the general theory of relativity which are capable of investigation, and to which the physics of pre-relativity days does not also lead, and this despite the profound difference in the fundamental assumptions of the two theories. In what follows, we shall again consider these important deductions, and we shall also discuss the empirical evidence appertaining to them which has hitherto been obtained.

(*a*) MOTION OF THE PERIHELION OF MERCURY

According to Newtonian mechanics and Newton's law of gravitation, a planet which is revolving round the sun would describe an ellipse round the latter, or, more correctly, round the common centre of gravity of the sun and the planet. In such a system, the sun, or the common centre of gravity, lies in one of the foci of the orbital ellipse in such a manner that, in the course of a planet-year, the distance sun-planet grows from a minimum to a maximum, and then decreases again to a minimum. If instead of Newton's law we insert a somewhat different law of attraction into the calculation, we find that, according to this new law, the motion would still take place in such a manner that the distance sun-planet exhibits periodic variations; but in this case the angle described by the line joining sun and planet during such a period (from perihelion—closest proximity to the sun—to perihelion) would differ from 360°. The line of the orbit would not then be a closed one, but in the course of time it would fill up an annular part of the orbital plane, viz. between the circle of least and the circle of greatest distance of the planet from the sun.

According also to the general theory of relativity, which differs of course from the theory of Newton, a small variation from the Newton-Kepler motion of a planet in its orbit should take place, and in such a way, that the angle described by the radius sun-planet between one perihelion and the next should exceed that corresponding to one complete revolution by an amount given by

$$+\frac{24\pi^3 a^2}{T^2 c^2 (1-e^2)}\,.$$

(*N.B.*—One complete revolution corresponds to the angle 2π in the absolute angular measure customary in physics, and the above expression gives the amount by which the radius sun-planet exceeds this angle during the interval between one perihelion and the next.) In this expression a represents the major semi-axis of the ellipse, e its eccentricity, c the velocity of light, and T the period of revolution of the planet. Our result may also be stated as follows: According to the general theory of relativity, the major axis of the ellipse rotates round the sun in the same sense as the orbital motion of the planet. Theory requires that this rotation should amount to 43 seconds of arc per century for the planet Mercury, but for the other planets of our solar system its magnitude should be so small that it would necessarily escape detection.[1]

In point of fact, astronomers have found that the theory of Newton does not suffice to calculate the observed motion of Mercury with an exactness corresponding to that of the delicacy of observation attainable at the present time. After taking account of all the disturbing influences exerted on Mercury by the remaining planets, it was found (Leverrier—1859—and Newcomb—1895) that an unexplained perihelial movement of the orbit of Mercury remained over, the amount of which does not differ sensibly from the above-mentioned + 43 seconds of arc per century. The uncertainty of the empirical result amounts to a few seconds only.

(b) DEFLECTION OF LIGHT BY A GRAVITATIONAL FIELD

In Section 22 it has been already mentioned that, according to the general theory of relativity, a ray of

1. Especially since the next planet Venus has an orbit that is almost an exact circle, which makes it more difficult to locate the perihelion with precision.

light will experience a curvature of its path when passing through a gravitational field, this curvature being
similar to that experienced by the path of a body which
is projected through a gravitational field. As a result of
this theory, we should expect that a ray of light which is
passing close to a heavenly body would be deviated
towards the latter. For a ray of light which passes the sun
at a distance of Δ sun-radii from its centre, the angle of
deflection (*a*) should amount to

$$a = \frac{1.7 \text{ seconds of arc}}{\Delta} .$$

It may be added that, according to the theory, half of
this deflection is produced by
the Newtonian field of attraction of the sun, and the other
half by the geometrical modification ("curvature") of space
caused by the sun.

Fig. 5

This result admits of an
experimental test by means of
the photographic registration
of stars during a total eclipse of
the sun. The only reason why we must wait for a total
eclipse is because at every other time the atmosphere is

so strongly illuminated by the light from the sun that the stars situated near the sun's disc are invisible. The predicted effect can be seen clearly from the accompanying diagram. If the sun (S) were not present, a star which is practically infinitely distant would be seen in the direction D_1, as observed from the earth. But as a consequence of the deflection of light from the star by the sun, the star will be seen in the direction D_2, *i.e.* at a somewhat greater distance from the centre of the sun than corresponds to its real position.

In practice, the question is tested in the following way. The stars in the neighbourhood of the sun are photographed during a solar eclipse.

In addition, a second photograph of the same stars is taken when the sun is situated at another position in the sky, *i.e.* a few months earlier or later. As compared with the standard photograph, the positions of the stars on the eclipse-photograph ought to appear displaced radially outwards (away from the centre of the sun) by an amount corresponding to the angle a.

We are indebted to the Royal Society and to Royal Astronomical Society for the investigation of this important deduction. Undaunted by the war and by difficulties of both a material and a psychological nature aroused by the war, these societies equipped two expeditions—to Sobral (Brazil) and to the island of Principe (West Africa)—and sent several of Britain's most celebrated astronomers (Eddington, Cottingham, Crommelin, Davidson), in order to obtain photographs of the solar eclipse of 29th May, 1919. The relative descrepancies to be expected between the stellar photographs obtained during the eclipse and the comparison photographs amounted to a few hundredths of a millimetre only. Thus great accuracy was necessary in making the adjustments required for the taking of the photographs, and in their subsequent measurement.

The results of the measurements confirmed the theory in a thoroughly satisfactory manner. The rectangular components of the observed and of the calculated deviations of the stars (in seconds of arc) are set forth in the following table of results:

Number of the Star.			First Co-ordinate.		Second Co-ordinate.	
			Observed.	Calculated.	Observed.	Calculated.
11	.	.	−0.19	−0.22	+0.16	+0.02
5	.	.	+0.29	+0.31	−0.46	−0.43
4	.	.	+0.11	+0.10	+0.83	+0.74
3	.	.	+0.20	+0.12	+1.00	+0.87
6	.	.	+0.10	+0.04	+0.57	+0.40
10	.	.	−0.08	+0.09	+0.35	+0.32
2	.	.	+0.95	+0.85	−0.27	−0.09

(*c*) DISPLACEMENT OF SPECTRAL
LINES TOWARDS THE RED

In Section 23 it has been shown that in a system K' which is in rotation with regard to a Galilean system K, clocks of identical construction and which are considered at rest with respect to the rotating reference-body, go at rates which are dependent on the positions of the clocks. We shall now examine this dependence quantitatively. A clock, which is situated at a distance r from the centre of the disc, has a velocity relative to K which is given by

$$v = \omega r,$$

where ω represents the velocity of rotation of the disc K' with respect to K. If v_0 represents the number of ticks of the clock per unit time ("rate" of the clock) relative to K when the clock is at rest, then the "rate" of the clock (v) when it is moving relative to K with a velocity v, but at rest with respect to the disc, will, in accordance with Section 12, be given by

$$v = v_0 \sqrt{1 - \frac{v^2}{c^2}},$$

or with sufficient accuracy by

$$v = v_0 \left(1 - \frac{1}{2} \frac{v^2}{c^2} \right).$$

This expression may also be stated in the following form:

$$v = v_0 \left(1 - \frac{1}{c^2} \frac{\omega^2 r^2}{2} \right).$$

If we represent the difference of potential of the centrifugal force between the position of the clock and the centre of the disc by ϕ, *i.e.* the work, considered negatively, which must be performed on the unit of mass against the centrifugal force in order to transport it from the position of the clock on the rotating disc to the centre of the disc, then we have

$$\phi = -\frac{\omega^2 r^2}{2}.$$

From this it follows that

$$v = v_0 \left(1 + \frac{\phi}{c^2} \right).$$

In the first place, we see from this expression that two clocks of identical construction will go at different rates when situated at different distances from the centre of the disc. This result is also valid from the standpoint of an observer who is rotating with the disc.

Now, as judged from the disc, the latter is in a gravitational field of potential ϕ, hence the result we have obtained will hold quite generally for gravitational fields. Furthermore, we can regard an atom which is emitting spectral lines as a clock, so that the following statement will hold:

An atom absorbs or emits light of a frequency which is dependent on the potential of the gravitational field in which it is situated.

The frequency of an atom situated on the surface of a heavenly body will be somewhat less than the frequency of an atom of the same element which is situated in free space (or on the surface of a smaller celestial body).

Now $\phi = - K (M/r)$, where K is Newton's constant of gravitation, and M is the mass of the heavenly body. Thus a displacement towards the red ought to take place for spectral lines produced at the surface of stars as compared with the spectral lines of the same element produced at the surface of the earth, the amount of this displacement being

$$\frac{v_0 - v}{v_0} = \frac{K}{c^2} \frac{M}{r} .$$

For the sun, the displacement towards the red predicted by theory amounts to about two millionths of the wave-length. A trustworthy calculation is not possible in the case of the stars, because in general neither the mass M nor the radius r is known.

It is an open question whether or not this effect exists, and at the present time astronomers are working with great zeal towards the solution. Owing to the smallness of the effect in the case of the sun, it is difficult to form an opinion as to its existence. Whereas Grebe and Bachem (Bonn), as a result of their own measurements and those of Evershed and Schwarzschild on the cyanogen bands, have placed the existence of the effect almost beyond doubt, other investigators, particularly St. John, have been led to the opposite opinion in consequence of their measurements.

Mean displacements of lines towards the less refrangible end of the spectrum are certainly revealed by statistical investigations of the fixed stars; but up to the present the examination of the available data does not allow of any definite decision being arrived at, as to whether or not these displacements are to be referred in reality to the effect of gravitation. The results of observation have been collected together, and discussed in detail from the standpoint of the question which has been engaging our attention here, in a paper by E. Freundlich entitled "Zur Prüfung der allgemeinen Relativitäts-Theorie" *(Die Naturwissenschaften,* 1919, No. 35, p. 520: Julius Springer, Berlin).

At all events, a definite decision will be reached during the next few years. If the displacement of spectral lines towards the red by the gravitational potential does not exist, then the general theory of relativity will be untenable. On the other hand, if the cause of the displacement of spectral lines be definitely traced to the gravitational potential, then the study of this displacement will furnish us with important information as to the mass of the heavenly bodies.

COMMENTARY

Robert Geroch

PREFACE

Relativity, despite its reputation, is not a particularly difficult subject. Many universities offer courses in special and general relativity to freshmen and sophomore non-science majors. Many hundreds of physicists throughout the world, including graduate students, carry out research in relativity on a day-to-day basis.

One good way to get a feeling for what relativity theory is all about is to read, in these pages, what the originator of the subject had to say. I have provided eleven comments, attached to various of the thirty-two sections of Einstein's book. Some explain in more detail the issue Einstein is raising; some describe more recent developments. The key to understanding relativity is to think about it for yourself. I hope these comments will serve that end.

Here is a brief outline of how this book is structured.

In Part I, consisting of seventeen sections, Einstein introduces the special theory of relativity. This theory is concerned primarily with the structure of space and time. Light serves as an essential tool to probe space and time, and so light itself comes to play an important role in the theory. Indeed, it is a mystery about how light propagates that sets the stage for the theory; and the final theory has a great deal to say about the behavior of light.

In Sections 1–3, Einstein introduces several notions that will play a role in the subsequent development. Among these are the idea of geometry, and the relation between geometry and the physical world. Nothing in these three sections bears directly on the theory of relativity. Rather, this material serves as a warm-up—as background for what will follow.

A key section is Section 5, dealing with the principle of relativity. This principle asserts, essentially, that one group of observers, occupying an inertial frame, will determine precisely the same laws of physics as some other group of observers, moving past the first group at constant velocity. This principle is of fundamental importance for the theory of relativity. It is well worth taking the time to think about and to understand it. The section preceding this one, Section 4, serves primarily to provide one example of a "law of physics" to which the principle of relativity can be applied. The section that succeeds it, Section 6, describes further the physics of life in an inertial frame.

The other key section of Part I is Section 7, on the law of propagation of light. What is found, in a nutshell, is that the way light propagates through space appears to be incompatible with the principle of relativity. It is this tension that gives rise to special relativity. If you feel relaxed and comfortable after you have finished reading Section 7, then you should read it (and perhaps also Section 5) again!

By the end of Section 7, Einstein has spelled out the apparent incompatibility that is at the heart of relativity. The remaining ten sections of Part I describe how this issue is resolved, i.e., what special relativity is and what its implications are.

In Sections 8 and 9, Einstein argues that any reconciliation of the principle of relativity and the observed law of propagation of light seems to require that the notion of "simultaneity" for two events must depend on who is looking at them. And, in Section 10, he does the same thing for the "spatial distance" between two events. Here is where we begin to explore the structure of space and time. This material is difficult because it goes against our intuition. It is important to understand that the ideas introduced in Sections 8–10 arise naturally from an attempt to reconcile the principle of relativity and the observed behavior of light. We seek the simplest—not the most esoteric—reconciliation. You don't have to accept what Einstein says in these three sections—especially if you can provide some alternative reconciliation!

What was only painted in broad brushstrokes in Sections 8–10 is made more precise in Section 12. This section contains the formulae that describe in detail how different observers will perceive the passage of time and spatial distances. The following section (Section 11), giving the formulae for the Lorentz transformation, is quite technical, and is perhaps best skipped on first reading.

Section 14 is a brief summary of the final form that the principle of relativity takes in light of the special theory.

Sections 13, 15 and 16 deal with a number of implications of special relativity. The Michaelson-Morley experiment (Section 16) is closely tied with our understanding of the propagation of light, as discussed in

Section 7, and should be read in connection with that section. The remaining material in these three sections serves to round out the theory: It may be helpful as an indication of how special relativity works, but is not essential to an understanding of the theory itself. The theoretical implications include the formula for combining velocities (Section 13), and for the conversion between energy and mass (Section 15). The experimental implications include the Fizeau experiment (Section 13) and the Michaelson-Morley experiment. That both of these experiments involve the propagation of light is no surprise, for the behavior of light lies at the heart of special relativity.

Finally, Section 17, on Minkowski's four-dimensional space(-time), plays two roles. On the one hand, it serves as a kind of wrap-up of the structure of space and time as described by the special theory of relativity. And, on the other, it serves as an introduction to the way relativity theory is viewed today.

In Part II, consisting of twelve sections, Einstein introduces the general theory of relativity. This is, in essence, a theory of the gravitational field. It turns out that gravitation is intimately connected with the geometry of space and time; and so the general theory becomes also a theory of space-time geometry.

Sections 18–21 introduce what is today called the principle of equivalence. The idea here is that, in a small region of space-time, the physical effects of a gravitational field are identical in every respect to the effects of acceleration. The key section is Section 20, in which the principle itself is formulated. This principle, in turn, rests heavily on the equality of inertial and gravitational mass, which is discussed in Section 19. Indeed, as Einstein points out, this equality can be viewed as a consequence of the equivalence principle. Sections 18 and 21 contain some ideas that serve as background for this principle, and expand on its scope. These sections introduce the idea of a non-inertial frame of reference; that such frames should be taken seriously; and that non-inertial frames might be linked to gravitational phenomena.

Sections 22 and 23, for general relativity, are analogous to Sections 8–10 for special relativity. They deal with some physical implications of the equivalence principle. Once we have decided that the effects of gravitation can be understood by looking at effects that occur in non-inertial frames, then we can make use of such frames to understand gravity.

Acceleration affects the behavior of measuring rods and clocks. So, by the discussion of Sections 22 and 23, gravitational fields will also affect measuring rods and clocks. But rods and clocks are the fundamental tools we use to study the "geometry" of space-time. In Sections 24–29, this idea is carried to its logical conclusion. If a gravitational field distorts rods and clocks, and if these are the objects by means of which geometry is measured, then a gravitational field must also distort the geometry of space and time. This "distortion" manifests itself as curvature in space-time. These six sections introduce the idea of curvature, and its connection with gravity.

Part III, consisting of three sections, introduces cosmology, the study of the universe as a whole. Our understanding of our universe has changed dramatically, primarily as a result of more recent experiments, since this book was written. As a result, some of the discussion in these sections is out of date. But the treatment of what observers in a closed universe would experience, in Sections 31 and 32, provides a fine overview of how the geometry of space-time is connected with its physics.

Appendix 1 (a derivation of the formulae for the Lorentz transformation) is probably too technical for the general reader. The imaginary time, introduced in Appendix 2, is now regarded as a less useful tool for providing insight into the structure of space-time. The reader may wish to skip these two appendices on first reading. Appendix 3 discusses the three famous "experimental tests" of the general theory of relativity.

SECTION 5. THE PRINCIPLE OF RELATIVITY

The principle of relativity is one of the important ideas underlying the special theory.

We set up our own frame of reference (say that of the room we now occupy); and we staff it by various observers. These observers examine various natural phenomena taking place in the room, and ultimately formulate an exhaustive description of what they observe. This description gives rise to what we call the laws of physics, as formulated in that frame. Now let there be introduced a second frame of reference, traveling by ours at constant velocity. That second frame is also staffed by observers, who also examine various natural phenomena, formulate a description of what they observe, and, ultimately, formulate the laws of physics in their frame. The principle of relativity asserts that the two sets of laws of physics—as formulated in the two frames—will in fact be the same.

It is important to note that the principle of relativity refers *only* to "laws of physics." Here are two examples of such laws: (1) that a body, acted upon by no external force, will always continue to move in a straight line, and (2) that heat always flows from a hotter body to a colder one. The principle of relativity, then, asserts that if observers in one frame of reference find these laws to hold, then observers in any other frame of reference will also find them to hold. But there are numerous other types of natural phenomena that would not qualify as "laws of physics" at all. Examples of such phenomena include the sight of a solar eclipse, or the sound of the explosion of a volcano. The principle of relativity does *not* assert that such phenomena will be seen identically in the two frames of reference. In the case of the volcano, for example, those in a frame of reference moving toward the volcano will hear the sound at a higher pitch than those moving away.

This distinction between what is and what is not a "law of physics" seems quite clear in the examples above. The eclipse and the volcano are "specific phenomena in the environment"; laws of physics, by contrast, are general, overriding principles that apply to a whole variety of specific phenomena. But there are other examples, in which the distinction is less clear-cut. For instance, those in a frame of reference at rest in the Earth's atmosphere will discover the following law of sound: "Sound travels, in all directions, at a fixed speed, about 1000 feet per second." Were this observation elevated to a "law of physics," then we would have a violation of the principle of relativity, for those in other frames of reference will not find this law to hold. They will find that the speed of sound is smaller in some directions (i.e., the direction of their travel through the atmosphere), and greater in others. In order to rescue the principle of relativity in this example, we must make the law of physics read: "Sound travels, in all directions, at a fixed speed, about 1000 feet per second, *relative to the underlying air*." In other words, we view the Earth's atmosphere as merely another feature of the environment, not unlike an eclipse of the Sun. This viewpoint is supported by the fact that there are non-sound ways of observing the air, e.g., by experiencing the wind. The "general, overriding principle," then, refers to the behavior of sound relative to the air. But in other examples, this distinction becomes still trickier. Our entire universe is bathed in low-level electromagnetic radiation, the "black-body radiation." Shall we think of this radiation—which is everywhere, and in the only universe we have—as merely a "specific phenomenon in the environment"?

The principle of relativity, then, hides within itself a subtle distinction—between what is and what is not taken as a law of physics. Indeed, it could be argued that a better perspective is to regard the principle of relativity, not as a general principle of nature at all, but rather as a guideline for distinguishing between those phenomena that are to be taken as "laws of physics" and those that are not. Phenomena that have the same description in every frame—that is, phenomena that are compatible with the principle of relativity—are to be accorded the status of physical laws, while phenomena that have different descriptions in different frames are to be regarded as merely specific phenomena. This is not a purely philosophical distinction: It can have consequences as to how physics is conducted. Suppose, for example, that subsequent experiments showed that sometimes, in certain reference frames, heat actually flows from a colder to a hotter body. Physicists would probably respond to such experiments, not by giving up the principle of relativity, but, rather, by searching for some "cause" for these observations—for some external, environmental factors that would account for what we are observing.

SECTION 7. THE PROPAGATION OF LIGHT

The key to understanding special relativity is understanding the behavior of light. Let us consider this behavior in more detail.

Light, on the one hand, could behave as fish swimming in the sea. A fish, a moment after its release from a slow-moving boat, swims at a speed independent of that of the boat. After all, it is merely swimming in the ocean at its normal fish-swimming speed. (Let us take all of these fish to be equally competent swimmers.) Suppose that an individual riding in this boat were to release various fish in various directions. This individual will see different fish moving at different speeds, depending on the fish's direction of travel. For example, a fish swimming in the direction in which the boat is traveling will be seen, by those in the boat, to be moving more slowly than one swimming in the opposite direction.

On the other hand, light could behave as bullets fired from a gun. A bullet, shot into the air from a moving boat, acquires the boat's speed. Let an individual, again riding in the boat, fire bullets in various directions. This individual will see all those bullets traveling away from him with the same speed (namely, the muzzle speed of the gun).

These two models differ from each other in other important respects. In the bullet model, for example, each bullet retains some "memory" (recorded in its motion) of how it originated. The various bullets, fired from various boats, are not all the same. For instance, one bullet could overtake another if, say, the first was fired from a faster-moving boat. There are, if you like, various "varieties" of bullets, corresponding to the varieties of motions of the boats from which the bullets were fired. In the fish model, by contrast, a fish has no memory of the boat from which it was released. No fish could ever overtake another. There is only a single variety of fish.

Here is another difference between these two models. Let us now impose a restriction on the individuals riding in their boats: They can observe only fish or bullets, and nothing else. That is, they are prohibited from watching the ocean, feeling the wind, etc. Under this prohibition, such individuals would not be able, by observing the bullets they fired, to determine whether theirs was a fast- or slow-moving boat. In both cases they would see all their bullets traveling away at some fixed (muzzle) speed. In other words, all the various boats in the bullet model are "equivalent" to each other.

Not so in the fish model. The individuals who release fish under the same restrictions *will* be able to tell whether they are riding in a fast- or slow-moving boat. For instance, in the case of a *very* slow-moving boat, all the released fish will appear to be moving away from the boat at approximately the same speed. But in the case of a fast-moving boat, the fish will appear to move at quite different speeds, depending on their direction. In short, the various boats in the fish model are not equivalent to each other.

So here, in any case, are two models for the behavior of light. In the fish model, there is a single type of projectile (fish), but a whole variety of observers (boats), which are demonstrably different from each other. In the bullet model, there is a whole variety of projectiles (bullets), but a single type of observer.

Which model correctly describes light? The answer, remarkably enough, is neither! Consider observers, in various states of motion, emitting light projectiles (flashes of light). These flashes of light behave like fish in that a light flash, once emitted, loses all information about its source. There is only a single type of light flash. No light flash, for instance, could overtake another. On the other hand, light flashes behave like bullets in that they do not distinguish between different observers. An observer, upon emitting light flashes in various directions, sees all those flashes moving away at some fixed speed (the "muzzle speed of light"). Thus, light pulses share some features of swimming fish, and some features of fired bullets.

It should be emphasized that these features of light are found in concrete physical experiments: They are not merely the fruits of some fancy theory or abstract argument. The bullet-behavior of light—the inability to use light to detect absolute motion—is seen in the Michaelson-Morley experiment. In this experiment, in essence, we emit light in different directions, and measure its speed. We find that the light speed is the same no matter what the direction of travel of the light. Furthermore, this holds for different motions of the observer who is emitting the light. The fish-behavior of light—that the motion of light is independent of that of its source—is seen by observation of double stars. A double star is a pair of stars in orbit about each other, which we observe by seeing the light that the stars emit. If the light acquired the speed of the star that emitted it, then the light emitted from the star moving toward us in its orbit would arrive earlier than the light from the star moving away from us. This would result in distortion in the appearance of the double-star system. But we see no such distortion, suggesting that a light pulse, once emitted, travels in a manner independent of the motion of the emitter.

So here, in a nutshell, is the problem that gives rise to relativity. We can imagine swimming fish, and we can imagine fired bullets—but it is very difficult to imagine projectiles combining the features of the two. Relativity, if you like, is merely a mechanism by which this particular combination of features is reconciled. Later on in this book, you may get the feeling that relativity is contradictory, or all too fantastic, or otherwise unacceptable. If and when this feeling arises, it might be worthwhile to return to this chapter. Your burden, if you wish to reject relativity, is to provide some alternative reconciliation of these two features of light—features that, remember, are seen in actual experiments. Find a reconciliation that is less "contradictory," less fantastic, or more acceptable. Such a reconciliation would, perhaps, be the seeds of an alternative to relativity theory.

SECTION 10. RELATIVITY AND INTUITION

This chapter, as well as the previous one, asks us to contemplate that the physical world might behave in a way that, at best, is counter to our intuition; and, at worst, is contradictory or just plain absurd. You might be convinced that two events are either simultaneous or they are not; that they are either separated by three feet or they are not. It seems so obvious! It defies common sense to imagine that these notions could depend on the observer.

Being able to entertain, at least in a tentative way, what at first seems absurd is a skill—one that can be acquired and then refined. Such a skill is undoubtedly useful in physics, where it helps us expand our horizons. Indeed, later in this book you will be asked to entertain the possibility that space is non-Euclidean, and that it is finite in extent. Quantum mechanics, to take a second example, asks us to entertain the possibility that "position" is not even an attribute of a particle. This skill is, arguably, useful in other areas of life as well.

Here is a training exercise, designed to hone that skill. Imagine you are having a discussion with a friend who, having been raised by wolves, is adamant in his belief that, say, the Earth is flat. This person will raise a number of objections to a round Earth. "If the Earth were round, then the people on the other side would feel very differently since they are all upside down. But I've never met anybody who claims to have lived upside down." "Since the people on the other side couldn't see the Sun when we do, their days and nights would be reversed. But nobody I know has experienced such a reversal." "If the Earth were round, the people on the other side would fall off." Your goal is to convince this person, not that their beliefs are wrong or silly, but rather that an alternative—a round Earth—is viable. That is, you wish to argue that a round Earth is not ridiculous on its face. The nature of your responses will, in many cases, be that your friend's intuition represents an extrapolation from his or her experience, that that extrapolation bridges numerous gaps in your friend's knowledge, and that there are, conceivably, other ways to traverse those gaps.

In any case, you can then take whatever you have said to your friend, adapt it to the claims of relativity, and say it back to yourself.

Here are some other counter-intuitive positions, on which you might try this exercise: (1) many diseases are caused by tiny living creatures that actually enter the body and interfere with its functioning; (2) people, in their present form, have arisen by a process, beginning with one-celled animals, through myriad small changes; and (3) the stars we see at night are actually large, bright objects, not unlike our Sun, but very far away. Indeed, should you eventually find yourself accepting the tenets of relativity, you may discover that it is not even the most absurd body of ideas to which you subscribe.

SECTION 12. LENGTH

The conclusion of this section—for the case of the length of a rod—may be summarized as follows. Take a rod, place a group of observers in its rest frame, and let them measure its length. Let observers in another frame, moving along the rod at some speed v, also measure its length. Then, according to relativity, those in the second frame will find the rod to be shorter, by the famous factor $\sqrt{1 - v^2/c^2}$. Note that when v is much smaller than the speed of light c, then v^2/c^2 will be a very small number indeed, and so the factor $\sqrt{1 - v^2/c^2}$ will be nearly one. For bullets and the like, v will be perhaps one millionth of c, v^2/c^2 will be about a million-millionth, and the factor will be about 0.9999999999995—hardly a shrinkage-factor one would notice on casual observation. In this sense, then, special relativity is in accord with our everyday observations.

It is important to understand that these effects are predicted to occur *only* if the length is measured in the specific manner that Einstein has in mind. Namely, the following. The individuals in the frame in which the rod is moving by must station two observers in the following manner. They must be so stationed that, just as the front end of the rod passes one of them, the back end of the rod is passing the other. The critical element here is that these two passages take place at the same instant of time: It will take a little planning to arrange the observers in precisely the correct positions. Then, after the rod has passed, the two observers must measure the spatial distance between them. They deem this distance to be the length of the rod. In the instructions above, "at the same instant of time" and "spatial distance between them" refer, of course, to the frame that these observers occupy, i.e., that in which the rod is moving by.

There are many other methods to measure "the length of a moving rod." For example, a single observer in this frame could determine the time it takes a light ray, beginning at one end of the rod, to traverse the rod, reach the far end, and then return to the other end. We could then deem the length of the rod to be half this elapsed time multiplied by the speed of light. Or, we could arrange for flashes of light to take place simultaneously on the two ends of the rod and determine the time required for those flashes to meet. A final alternative would be for an observer, stationed far from the rod, simply to look at the rod as it moves by, noting whether it appears to have shortened.

Now, in the case of a rod sitting on a table at rest in the laboratory, these various methods yield the same answer. We think of them as merely different methods of measuring the same thing, the length of the rod. Indeed, the reason why we have a single word, "length," to describe all these different measurement methods is precisely this belief that all these methods *must* yield the same answer. But the realm of relativity—with objects moving by at speeds comparable with the speed of light—is very much outside of our everyday observations. There is considerably less reason to believe that these various methods must all agree in this realm. In fact, according to relativity they do not. For each different method there will be a factor—which can be calculated within special relativity—for what happens to the length of the rod when measured by that method. For one method that factor might be one, i.e., that method might always yield precisely the rest-length of the rod; and for some other method that factor might be greater than one, i.e., the rod, as measured by that method, would be deemed to have lengthened rather than shortened.

Thus, "moving objects contract" is not a very good summary of what is predicted by special relativity. Suppose, for a moment, that relativity was within the realm of our everyday observations, e.g., that typical automobile speed limits were half the speed of light. In this world, the word "length" in our language would probably have been replaced by a great number of different words, corresponding to the great number of different notions of "length" that we would experience.

Similar remarks apply to "time dilation." There are numerous experiments the results of which deserve to be called the "elapsed time" between two events; and different experiments will generally yield different answers. Einstein's time dilation refers to one particular choice of an experiment.

SECTION 15. THE CONVERSION OF ENERGY TO MASS

The "equivalence of mass and energy"—represented by the famous equation $E = mc^2$—has certainly captured the popular imagination. It has been credited with providing a limitless source of useful energy for society, and blamed for the existence of nuclear weapons. But, arguably, this little equation deserves neither the credit nor the blame it receives.

This "equivalence" is, in principle, a very simple idea. Take a brick, and weigh it. Then heat the brick in an oven, being very careful neither to let it lose matter (e.g., by evaporation of surface moisture) nor to gain matter (e.g., by deposit of soot). Now weigh the brick again. This equation asserts that the hot brick will weigh more than the cold brick did, since energy (in the form of heat) was transferred to the brick. In a similar vein, a pinball machine with the balls rolling about would weigh more (by virtue of the energy of ball motion) than would the same machine with the balls at rest; a fully charged automobile battery would weigh more (by virtue of electrical energy) than the same battery when discharged.

It would seem, given that there are such simple examples of how this equation operates, that it would be easy to test it experimentally. Unfortunately, this is not the case, because the effects are very small indeed. A five-pound brick, baked in a 350° oven, would gain in mass about one ten-billionth of an ounce! Such a small mass change would be very hard to detect, not least because, as anyone who has ever baked knows, it is very difficult to avoid spurious mass gains and losses—which typically involve much more than one ten-billionth of an ounce—in the baking process.

There is good experimental evidence for the equation $E = mc^2$, but it is of a less concrete nature. We can measure the masses of nuclei and elementary particles by watching how they behave when pushed about by electric and magnetic fields; and we can measure the energy released when reactions take place in which one type of particle is converted to another. Thus, we can determine the mass lost in a reaction (by measuring the masses before and after), and the energy released in that reaction. In this way, we measure both the E and the m that appear in the equation, and we check the equation itself. In contrast to the brick example, up to a few percent of the original mass is converted into energy in these processes. Thus, we may check the equation $E = mc^2$ with reasonable accuracy.

Is $E = mc^2$ "responsible" for nuclear power, nuclear weapons, and the like? In essence, we are asking whether there would still be nuclear power if $E = mc^2$ failed. But the laws of physics are all interrelated with each other, making it very difficult to say what the world would be like if one such law were changed (and all the others remained the same). If Newton's laws of inertia were repealed, all others remaining the same, would it still be possible to play baseball? My sense is that it would be possible to invent a plausible-looking theory of nature in which $E = mc^2$ failed, but enough of the structure of nuclear physics remained intact to allow working nuclear power plants.

Here is another, perhaps more revealing, way to look at this issue. When a rabbit runs across a field, that rabbit is losing mass (because it is expending energy). In other words, the rabbit is converting mass into energy. We might well argue, therefore, that $E = mc^2$ is responsible for the ability of rabbits to run across fields.

***

The speed of light c (about 670,000,000 miles per hour) is large by everyday standards. Consequently, the equation $E = mc^2$ implies that a moderate amount of energy represents a very small change in mass. Exactly how small? Here is a simple way to understand the physical content of the equation $E = mc^2$.

Consider an object that acquires, in some way, some energy: A brick is heated up, or a battery is charged. We wish to know by what fraction the mass of this object is thereby increased. The rough answer can be determined as follows. Imagine that this energy, instead of being merely transferred to the body (as heat, electrical energy, or whatever), were instead used to move the body— to accelerate it up to some final speed. Thus, in the example of the brick, the heat energy might be used, not to heat the brick, but rather to drive a steam engine to propel the brick along a track. Take the final speed reached in this way, divide by the speed of light (the result being, typically, a very small number), and then square that number. The result is approximately the fractional mass-increase the body would suffer on absorbing that energy. Consider, as an example, the case of a battery being charged. Say that, had the charger instead been used to run an electric motor to propel the battery, then the battery would have reached a final speed of 670 miles per hour. (This is one millionth [a fraction 10^{-6}] of the speed of light.) Then, had we used this battery charger merely to charge the battery, the mass of the battery would be increased by about one part in a million-million (i.e., by a fraction $10^{-12} = 10^{-6^2}$).

There is another side of this coin. Since c is so large, a very small amount of mass is in principle capable of generating an enormous amount of energy. For example, the matter in a sofa, if converted entirely into energy, would satisfy the electric-power needs of the United States for a year. (Here, in light of the paragraph above, is another way to say the same thing: In order to get this sofa moving at an appreciable fraction of the speed of light, we would have to utilize the entire output of the electrical power grid of the United States for an entire year.) One might imagine that by now some inventor would have built a machine to do this, with the consequence that our electric rates would be much lower than they are. Unfortunately, there is a law of physics that appears to stand in the way of this project. The vast majority of the mass of a physical object is represented by the neutrons and protons that reside in the nuclei of its atoms. The law of baryon conservation asserts that neutrons and protons cannot be converted entirely into useful energy, but rather only into other neutrons or protons, or similar particles. This law, then, locks up—makes unavailable for conversion into energy—most of the mass of a normal object. You might wish it were otherwise; that we did not have to contend with baryon conservation; that we could unlock all the energy in even a few grains of sand. Be careful what you wish for. Without the law of baryon conservation, it is probable that the neutrons and protons in the universe would long ago have turned, spontaneously, into energy. As a result, there would be no neutrons or protons left, and so no atomic nuclei, no atoms, and no matter as we know it. Thus, we would have plenty of power for illumination today, but no cities to illuminate!

SECTION 17. SPACE-TIME

Minkowski's viewpoint represents a "geometrization" of relativity. These ideas have, over the years, come to the forefront: They reflect the perspective of the majority of physicists working in relativity today. Let us expand on this viewpoint.

The fundamental notion is that of an *event*, which we think of as a physical occurrence having negligibly small extension in both space and time. That is, an event is "small and quick," such as the explosion of a firecracker or the snapping of your fingers. Now consider the collection of all possible events in the universe—all events that have ever happened, all that are happening now, and all that will ever happen; here and elsewhere. This collection is called space-time. It is the arena in which physics takes place in relativity. The idea is to recast all statements about goings-on in the physical world into geometrical structures within this space-time. In a similar vein, you might begin the study of plane geometry by introducing the notion of a point (analogous to an event) and assembling all possible points into the plane (analogous to space-time). This plane is the arena for plane geometry, and each statement that is part of plane geometry is to be cast as geometrical structure within this plane.

This space-time is a once-and-for-all picture of the entire physical world. Nothing "happens" there; things just "are." A physical particle, for example, is described in the language of space-time by giving the locus of all events that occur "right at the particle." The result is a certain curve, or path, in space-time called the world-line of the particle. Don't think of the particle as "traversing" its world-line in the same sense that a train traverses its tracks. Rather, the world-line represents, once and for all, the entire life history of the particle, from its birth to its death. The collision of two particles, for example, would be represented geometrically by the intersection of their world-lines. The point of intersection—a point common to both curves; an event that is "right at" both particles—represents the event of their collision. In a similar way, more complicated physical goings-on—an experiment in particle physics, for example, or a football game—are incorporated into the fabric of space-time.

One example of "physical goings-on" is the reference frame that Einstein uses in his discussion of special relativity. How is this incorporated into space-time? The individuals within a particular reference frame assign four numbers, labeled x, y, z, t, to each event in space-time. The first three give the spatial location of the event according to these observers, the last the time of the event. These numbers completely and uniquely characterize the event. In geometrical terms, a frame of reference gives rise to a coordinate system on space-time. In a similar vein, in plane geometry a coordinate system assigns two numbers, x and y, to each point of the plane. These numbers completely and uniquely characterize that point. The statement "the plane is two-dimensional" means nothing more and nothing less than that precisely two numbers are required to locate each point in the plane. Similarly, "space-time is four-dimensional" means nothing more and nothing less than that precisely four numbers are required to locate each event in space-time. That is all there is to it! You now understand "four-dimensional space-time" as well as any physicist.

Note that the introduction of four-dimensional space-time does *not* say that space and time are "equivalent" or "indistinguishable." Clearly, space and time are subjectively different entities. But a rather subtle mixing of them occurs in special relativity, making it convenient to introduce this single entity, space-time.

In plane geometry, we may change coordinates, i.e., relabel the points. It is the same plane described in a different way (in that a given point is now represented by different numbers), just as the land represented by a map stays the same whether you use latitude/longitude or GPS coordinates. We can now determine formulae expressing the new coordinate-values for each point of the plane in terms of the old coordinate-values. Similarly, we may change coordinates in space-time, i.e., change the reference frame therein. And, again, we can determine formulae relating the new coordinate-values for each space-time event to the old coordinate-values for that event. This, from Minkowski's geometrical viewpoint, is the substance of the Lorentz-transformation formulae in Section 11.

A significant advantage of Minkowski's viewpoint is that it is particularly well-adapted also to the general theory of relativity. We shall return to this geometrical viewpoint in our discussion of Section 27.

SECTION 19. INERTIAL AND GRAVITATIONAL MASS

The principle of equality of inertial and gravitational mass is a cornerstone of general relativity. If this principle were found to be violated, then, as we shall see in the next section, the entire edifice of general relativity would fall.

The mass of a body represents the "quantity of matter" that it contains. But "quantity of matter" does not define itself: We have to spell out in detail the experiment we intend to perform. That is, we must choose some physical effect associated with the "quantity of matter" in a body, and design an experiment that is sensitive to that effect. It turns out that there are a number of possible such effects, resulting in a number of possible notions of mass. The situation here is quite similar to that of "length," as discussed in the comment on Section 12.

Let us now consider the principle of equality of inertial and gravitational mass. Take two objects—a brick and a pencil. Weigh them, and determine the ratio of their weights: Let's say the brick weighs 1000 times as much as the pencil. (In technical terms, the gravitational mass of the brick is 1000 times the gravitational mass of the pencil.) Next, arrange some fixed force—say that produced by a certain coiled spring compressed to a certain extent. Subject the brick to that fixed force for some fixed time, and determine the final speed the brick attains. Do the same for the pencil—using the same fixed force and the same fixed time—and determine its final speed. We expect, of course, that the pencil will reach a greater final speed, for it is lighter. The principle of equality of inertial and gravitational mass asserts that the pencil will reach a final speed 1000 times that of the brick, given that the brick weighed 1000 times as much as the pencil. In other words, this principle asserts that objects that respond more to gravity (larger gravitational mass) will, in the same proportion, resist being moved by a fixed force (larger inertial mass).

Essentially, this experiment has been carried out. Instead of a pencil and brick, two cylinders made of different metals were employed. For the gravitational field, that of the Sun was used (this field having the great advantage that its direction changes throughout the day). The gravitational and inertial masses of the two cylinders were compared by observing their reactions to this time-varying solar gravitational field. This experiment demonstrated the equality of inertial and gravitational mass to a remarkably high accuracy—a few parts in a million-million (about 10^{-11}).

Note what has happened here. What once was a single quantity, mass, becomes two quantities—inertial and gravitational mass—when we investigate the possible mass-measuring experiments in more detail. We then raise the issue of whether these two types of mass are indeed equal to each other. Interestingly enough, "gravitational mass" itself has, subsequently, been replaced by two types of gravitational mass. Passive gravitational mass describes the response of an object to an external gravitational field; while active gravitational mass describes the ability of an object to create a gravitational field. Thus, when you weigh yourself, you are detecting your own passive gravitational mass and the Earth's active gravitational mass. (Experiments have also been carried out to show equality of active and passive gravitational mass, but these are far less accurate than the experiment just described.) Passive gravitational mass was the subject of the discussion above; and it is the equality of passive gravitational mass and inertial mass that is important for general relativity.

SECTION 23. THE ROTATING DISK

There has been a certain amount of confusion surrounding Einstein's discussion of the rotating disk, not least because that discussion appears in Part II, "The General Theory of Relativity."

It is certainly true, as Einstein points out, that the individuals rotating with the disk will observe a number of intriguing physical effects. For example, they will, even though they remain "at rest," experience certain forces, and these forces will have an exotic distribution in space. Furthermore, these individuals will discover that their spatial geometry, as determined by using small rods that they place in various positions on the disk, is not a flat one. And finally, they will discover, using clocks they place in various positions on the disk, that, for them, there is no universal time. In short, these individuals will determine that they definitely are not living in one of the inertial frames of special relativity. They will not be able to label events by the coordinates x, y, z, t in the usual way—that is, with the first three giving "position" in a Euclidean geometry, and the last giving universal time.

Nevertheless, the discussion of the physics of a rotating frame is carried out completely within the framework of special relativity. This is possible because there is also available here an inertial frame—that in which the disk is seen to be rotating—in which special relativity *is* applicable. Within this inertial frame, we may describe completely the physics of this problem: the rotating disk itself; the observers on the disk; the little rods they use; their clocks; their calculations, notes, and discussion. We may determine what physical laws they will derive, the manner in which they will verify and interpret those laws, and so forth.

To summarize: Special relativity suffices to describe fully the physics of the rotating disk, including all effects the individuals rotating with the disk observe. Nothing in this treatment requires general relativity.

Then why not introduce the rotating disk in Part I? Einstein could certainly have done so; but he didn't, presumably because it would have seemed rather extraneous at that point of the discussion. Then why introduce the rotating disk at all? Here is the reason. Having completed our treatment of special relativity, we now wish to determine the laws of gravitation. The key guideline we have to help us do this is the idea that the effects of gravitation are indistinguishable from those of inertia. So, we seek physical situations, *within the context of special relativity*, in which interesting inertial effects are on display. Since special relativity has already been established, we can now use that theory to determine what physical effects will occur in these situations. In this manner, we may establish guidelines as to what physical effects are expected to occur in a gravitational field. The rotating disk is one such physical situation.

SECTION 27. SPACE-TIME AND GENERAL RELATIVITY

In Section 17, we discussed Minkowski's geometrical viewpoint as it applies to special relativity. Recall that the idea there was to introduce space-time, whose points correspond to physical events, and to agree to describe all physics in terms of various geometrical constructs on this space-time. In particular, each reference frame of special relativity gives rise to a certain coordinate system on space-time. Four coordinates are required to label an event in space-time: Space-time is four-dimensional. We remarked in that earlier discussion that this viewpoint is particularly well adapted to general relativity as well. Let us see, now, how this comes about.

We begin by returning to our earlier discussion of elementary geometry. Consider a non-planar surface, say that of a sphere. Once again, we may introduce as the fundamental notion a single point, and then agree to assemble all those points into the surface of interest. And, again, we may agree to use this surface as the arena in which various geometrical constructions are to be carried out. It is immaterial whether that surface is a plane, a sphere, or some other curved surface. Next, we decide to introduce coordinate systems. The key feature of such a system is that it allows us to label each point uniquely by numbers—the coordinate-values assigned to that point. In the present case, two coordinates are required: Our surface is two-dimensional. And finally, given two such coordinate systems, we could determine formulae expressing the new coordinate-values for each point in terms of the old coordinate-values for that point. Note that, through all this, the treatments of the plane, the sphere, and any other surface, are all identical.

The same circumstance holds with respect to space-time. We may agree—whether in special relativity or general relativity (or, indeed, in most other, competing, theories of gravity)—to introduce the event as the fundamental notion, and to assemble all possible events into the fabric of space-time. We agree to describe physics geometrically within the arena of this space-time. Thus, we still have the notions of the world-line of a particle, the geometrical characterization of the collision of two particles, etc. And we may agree to introduce coordinate systems on this space-time, subject only to the condition that the coordinate-values uniquely characterize the event. Four numbers are required: Space-time is four-dimensional. Here is an (unorthodox) example of an allowed coordinate system in space-time.

We hire four pilots to fly about the room, each carrying a watch. We don't care whether the watches keep "good time": Their ticking rate may vary, depending on various external or internal conditions. Nor do we care whether the pilots fly in straight lines, or perform loops, or whatever. Now let there be marked an event in the room, say the explosion of a firecracker. Each pilot writes down the time at which he hears the explosion, as read by his own watch. We do not care whether or not the path of the sound through the air is distorted, e.g., by a wind that might be blowing through the room. In any case, these four numbers—each written down by one of the four pilots—will in general uniquely describe the event originally marked by the firecracker. This is a coordinate system. Clearly, there is an enormous variety of possible coordinate systems here. Given any two such systems, we could determine formulae, expressing the new coordinate-values for each event in terms of the old coordinate-values for that event. All of these considerations—the entire geometrical viewpoint up to this point—are the same for special relativity as for general relativity.

Returning to elementary geometry, recall, from Section 24, that individuals using small measuring rods within their surface are able to determine distances: Given any two nearby points of the surface, they may measure the distance between those points. These distances—for *every* pair of nearby points—constitute the "geometry" of their surface. Thus, we may think of the geometry as a very long list, where each entry in this list specifies two nearby points of the surface and then gives the measured distance between those points. Clearly, this list—the pattern of distances—will be quite different for the case of a sphere than for a plane. That is, these individuals may, by examining this geometry, determine whether their surface is flat (a plane) or curved (e.g., a sphere). This is the *first* point in this entire development at which there emerges any distinction between a flat and curved surface. Note that these individuals do not need to view the surface from the outside in order to make this distinction.

In the case of space-time, the small measuring rods are replaced by clocks, light rays, and, again, small measuring rods. (Extra instruments are needed because "time" has now also been included.) Instead of a pair of nearby points of the surface, we consider a pair of nearby events of space-time. Instead of the ordinary distance between two points on the surface, we measure what is called the "interval" between the two events, as discussed in Section 25. The "geometry" of space-time now again consists of a long list, each entry of which specifies two nearby events and then gives the measured interval between those events. Again, the pattern of intervals will be quite different for a flat space-time as for a curved space-time. Thus, these individuals may, by examining this geometry, determine whether their space-time is flat or curved. If space-time is flat, special relativity is applicable; if it is curved, general relativity is applicable. Here is the first point in this entire development in which this distinction comes into play. Note that there is no "outside" of space-time, from which we may get an external view of its curvature. Individuals within a two-dimensional surface who discover, by measuring its geometry, that their surface is in fact a plane have the option of introducing a particularly simple and natural class of coordinate systems: the Cartesian coordinates. This is an option: There is no requirement that they use such a system. If their geometry is curved, then this option is not available at all. Similarly, if observers in the physical world find that their space-time geometry is flat (special relativity), they have the option of introducing the coordinates, discussed in Section 17, associated with an inertial reference system. There is no requirement to use that system even in special relativity, as we saw in the comment on Section 23. When space-time is curved (general relativity), the option to introduce coordinates attached to an inertial reference system is not available.

This, then, is Minkowski's geometrical viewpoint of relativity. Its salient feature is that virtually the entire treatment is carried out in precisely the same way for special as for general relativity. It is only at the very end, when curvature comes into play, that this distinction becomes significant.

SECTION 28. SPECIAL RELATIVITY AND GENERAL RELATIVITY

It is perhaps appropriate at this point to discuss briefly how special and general relativity compare; and how these theories are viewed today.

According to Minkowski's geometrical viewpoint, as discussed in comments on Sections 17 and 27, both special relativity and general relativity start out in the same way. Each involves events, space-time as the collection of all events, a description of physics in the language of space-time, and a variety of coordinate systems that can be introduced onto space-time. The key difference arises when the geometry of this space-time is studied. If that geometry turns out to be flat, then special relativity is the relevant theory; if it is curved, then general relativity is relevant. Think of plane geometry as "special geometry," and of the geometry of a curved surface as "general geometry." In physical terms, the curvature of space-time is associated with gravity. So, general relativity is a theory of gravity, and special relativity is the theory that is applicable when gravitational fields do not play a significant role.

In Section 22, Einstein points out that one should not think of general relativity as "overthrowing" special relativity. The remarks above amplify Einstein's point. It is perhaps more accurate to think of special relativity as a particular, limiting case of general relativity—that in which the gravitational field is vanishingly small. In a similar vein, the geometry of a curved surface does not "overthrow" plane geometry. Plane geometry is a particular, limiting case of general geometry, where the limit in this case is that in which the curved surface approaches a plane.

Physicists working on relativity theory today tend to think of there being but a single theory of relativity—the general theory. Its structure is based on the geometrical viewpoint discussed above. This theory encompasses at the same time the structure of space and time, the behavior of light, and gravitational fields. The special case in which the gravitational field happens to be vanishingly small (the subject of Part I) is certainly of great interest, but it needn't be regarded as a separate theory in its own right. This situation is somewhat similar to that of Newton's laws of motion in mechanics. These laws are the basis for *the* theory of mechanics. They have as a special case the situation in which there is no force on a particle—and in which, as a consequence, that particle undergoes straight-line motion.

While this is certainly an important special case of general mechanics, it is regarded as merely that, and not a separate theory of "special mechanics."

Perhaps the words "special" and "general" are, therefore, somewhat unfortunate. These terms were certainly natural in the context of how Einstein originally thought about his theory. Einstein considers first how various inertial frames of reference—each moving past the other at constant velocity—are related to each other. Here the thrust is trying to understand the behavior of light. These are the "special" frames of reference, and their relation and structure is "special relativity." Next comes gravity. By the general principle of relativity, the effects of gravitation are indistinguishable from those of inertia. Thus, we are led, in order to study gravity, to consider frames more general than the inertial ones. The resulting theory of gravity is called "general relativity."

In short, there is but one theory of relativity, and its division into these two parts has more to do with its history than with its logical structure.

SECTION 32. COSMOLOGY

Our view of cosmology—of the structure of the universe as a whole—has changed radically since this book was written.

It is natural to think that our universe might have the feature called spatial homogeneity. This means that, on the average, one region of space, at any given epoch, is pretty much the same as any other region—that the universe does not change its character appreciably from one region to another. There is fairly good evidence for spatial homogeneity. When we look out at the distant galaxies, we find them to be fairly evenly distributed over the night sky. Such a uniform distribution is exactly what one would expect in a spatially homogeneous universe. Based on such evidence it was generally believed, already at the time this book appeared, that our universe was spatially homogeneous. This belief has, essentially, persisted to the present day. Indeed, there is additional evidence today in favor of this belief. It has been found that our universe is bathed in low-level electromagnetic radiation—called the "black-body radiation"—which is understood to have arisen in the early stages of the formation of the universe. But this radiation, too, is seen to come to us with approximately equal strength in all directions; behavior that, again, would be expected in a spatially homogeneous universe. But what of the dynamics of the universe? Evidence for the behavior of the universe over time is harder to come by, for we expect that, whatever it is that is happening in the universe, it is happening very slowly. But we have to start somewhere, and, lacking evidence to the contrary, a natural initial assumption was that our universe is static. This means that one epoch of time is pretty much the same as any other—that the universe does not change its character appreciably from one time to the next. Indeed, the assumptions that the universe is spatially homogeneous and that it is static fit together in a very nice way. Taken together, these two assumptions amount to assuming that our universe is homogeneous in *space-time* itself—that one region of space-time is pretty much the same as any other region of space-time.

So, let us make the tentative assumption that our universe is indeed static. We now wish to write down some model of our universe, within the structure of the general theory of relativity. This seems like a natural thing to do: General relativity is our theory of gravity, and gravity is certainly important to the structure of the universe. Alas, there turns out to arise a serious problem with this program: General relativity admits no models whatever for a static, spatially homogeneous universe containing galactic matter. It is not difficult to see physically why this should be so. The gravitational behavior of matter in general relativity is very similar to its behavior in the Newtonian theory of gravity: The gravitational force between material bodies is attractive. This means, with respect to our own universe, that each galaxy will be gravitationally attracted to all the others, with the result that the whole system will tend to "fall in on itself." But such behavior would violate the assumption that the universe is static.

So, we have a dilemma: Our assumption that our universe is static conflicts with what is permitted by general relativity. One way to rescue the situation is to change general relativity slightly. It is possible to modify the theory by introducing a new constant, called the "cosmological constant," whose physical effect is to impose, on a large scale, a small repulsive force. This force, if adjusted in just the right way, can be made to compensate precisely for the average gravitational attraction between the galaxies. Thus, we can indeed provide static models for our universe within general relativity, so modified. These models come in several varieties. Some are "finite but unbounded," as described by Einstein in this section; others are infinite and unbounded.

All this changed with the discovery that our universe is expanding. The evidence for this behavior was the observation that the light from distant galaxies suffers what is called a redshift—an indicator that that light was emitted from an object moving away from us. The expansion of our universe, it turns out, is completely compatible with spatial homogeneity. But an expanding universe is changing with time: For example, its mass density decreases with time, as the galaxies move further apart. That is, an expanding universe cannot be static. Thus, the motivation for the cosmological constant has suddenly disappeared. We may now return to general relativity as it was originally formulated, and, indeed, we find within that theory suitable models for our spatially homogeneous, expanding universe.

But, alas, this success was short-lived. Originally, it was possible to measure, essentially, only the current expansion rate of the universe. This single value was found to be consistent with the general-relativistic models. However, more recent measurements have given us more detailed information about this expansion. In particular, we can now measure not only the current rate of expansion of our universe, but also a second number, which represents, essentially, the rate at which that expansion is changing over time. These new, refined, data turn out to be inconsistent with *all* the spatially homogeneous models within general relativity as originally formulated. In order to achieve compatibility with these data, it has been necessary to go back and reintroduce the cosmological constant.

It should be emphasized that the process of gathering data in cosmology is a difficult one, and that these data are subject to change and reinterpretation as new techniques and ideas come into play. Our understanding of our universe is changing rapidly. Surely the final chapter on cosmology and general relativity has yet to be written.

THE CULTURAL LEGACY OF RELATIVITY THEORY

David C. Cassidy

The century since the publication of Einstein's first paper on relativity in 1905 has seen a profound and continuing transformation in the way we understand both humanity and the natural world. This transformation is driven by the power of human reason. It is evident in the advance toward a deeper understanding of the cosmos, the genetic code of life, and the elementary structure of matter, and in the myriad cultural benefits, technological devices, and financial fortunes such advances have inspired. But it was also a century that revealed the potential depths of human depravity in two world wars, totalitarian dictatorships, mechanized genocide, weapons of mass destruction, and now the scourge of terrorism.

Arriving in an era of such extremes, Einstein and relativity theory fostered a host of new discoveries, new modes of thinking about science and nature, and new levels of acceptance, rejection, and exploitation of research. The theory created a legacy that extended long after its initial publication and far beyond the narrow confines of physics. In the very broadest terms, that legacy may be seen in the varied responses to the problem of *meaning* so poignantly raised by relativity. This was a highly technical new theory that gave new meanings to familiar concepts and even to the nature of theory itself. The general public looked upon relativity as indicative of the seemingly incomprehensible modern era, educated non-scientists despaired of ever understanding what Einstein had done, and political ideologues used the new theory to exploit public fears and anxieties—all of which opened a rift between science and the broader culture that continues to expand today. At the same time, those enamored of science and relativity defended the theory from an anxious public. They tried to clarify its concepts, and, using relativity as a model, attempted to formalize the structure of scientific advances. Although such efforts have been highly influential, they too have led, in the end, to a further separation of the sciences from public culture, and to a further backlash against science in some quarters that continues a century later.

Einstein's aim in his four papers of 1905 was to obtain a unified foundation for physics. It was an ambitious undertaking, yet fully in line with the optimism of its age. Since the culmination of the Scientific Revolution in Newton's law of universal gravitation in the late 1600s, faith in the ability of human reason to illuminate the laws of nature—including human nature—pervaded nearly every branch of cultural and intellectual endeavor. This faith was reinforced during the decades preceding the new century by the success of electromagnetic theory and the technological explosion of the electric age. Joined with the so-called mechanical worldview on nature, these advances encouraged public optimism for a future filled with new triumphs of human reason, new labor-saving devices, new economic opportunities, and a new diffusion of democratic ideals. The researchers of natural laws became cultural, even moral, icons. Science, in partnership with business and government, promised to bring unlimited progress and prosperity to the industrialized world and beyond.

But within a decade of Einstein's first theory of relativity—the special theory—the world found itself shockingly mired in war. The spectacle of the most advanced cultural nations engaged for years in brutal trench warfare, starvation blockades, and even chemical weaponry, shook European faith in human progress led by the lamp of enlightened reason. The period following the world war brought even greater challenges through the upheavals of urban industrialization, the defeat and democratization of Germany, and bloody Communist and anarchist uprisings scattered across Europe and the United States in the wake of the Bolshevik Revolution. For large segments of the population, cynicism and a sense of lost meaning and direction suddenly replaced the naive optimism of the prewar years.

Against this uneasy background, Einstein and relativity theory appeared on the world stage as a reinforcement for some and a contradiction for others. On November 6, 1919, a British astronomical team reported that, during a total eclipse of the sun, the positions of stars near the sun had appeared to shift slightly from their known positions in close agreement with the predictions of Einstein's general theory of relativity, published in 1916. Scientists everywhere applauded Einstein's achievement. Overnight, Einstein became a mythical hero for stealing the secrets of nature through the power of reason in an era seemingly controlled by irrational forces. But for many, the bewildering new theory symbolized the irrational world in which they found themselves. While some artists and poets attempted to build the new understanding of nature into their works, others, such as Dadaist painter Raoul Hausmann, reveled in the sense of disruption. Hausmann's *Dada siegt* (Dada triumphs) (1920) shows a scientist with his brain exposed against a mishmash of mechanical objects, machinery, and a street scene, with the quotation emerging from his mouth, "The finer natural forces." Science, it seemed, was not only associated with, but responsible for, the triumph of meaningless chaos.

Relativity was not just another important new theory. It profoundly challenged the common understanding of everyday physical concepts—space, time, mass, simultaneity. Space and time were no longer what they had seemed; and mass was now strangely surrounded by a curved four-dimensional space-time world not envisioned in any high-school geometry text. Even the very name "theory of relativity," coming after the rise of Darwin's theory of evolution, seemed to confirm the decline of old absolute values and beliefs, together with the old world order, and the triumph of a universal *relativism*.

Einstein, of course, objected to such interpretations.

Relativity theory had nothing to do with relativism, he insisted. In fact, he had first called it the "theory of invariants," for its emphasis on the unchanging character of natural laws within different reference frames. The former physics teacher also invested considerable effort in popular accounts of the new theory, such as this one. And he argued vehemently against applications of scientific concepts to areas outside of science, declaring: "I believe that the present fashion of applying the axioms of physical science to human life is not only a mistake but has something reprehensible in it."(1)

Einstein's effort to unify physical science required, in its conclusions, that space, time, and mass exhibit different measured values, depending upon the speed of the measured object. Even though these and other effects become significant only at extremely high speeds (approaching the speed of light) or for events in the presence of enormous masses (stars and galaxies)— conditions not normally encountered in everyday life— people of all stations were puzzled and dismayed. Moreover, Einstein had not produced these results under the constraint of new experimental evidence demanding the construction of a new theory. Rather, he had worked essentially alone, in the quiet "temple of science," as he put it, starting from so-called freely invented universal hypotheses in a deductive theory whose results were compared only in the end with "sense experiences" for their "usefulness."(2) While the special theory did not require much unusual mathematics, the general theory required, beyond its fundamentals, a mathematical sophistication that even Einstein found difficult. However inaccurate, the striking portrait of a lonely, isolated genius spinning out an incomprehensible, abstract theory rendered Einstein and relativity obvious targets for public misunderstanding, misinterpretation, and misuse.

Although excellent popularizations of relativity existed early, and many non-science students today learn the fundamentals of the theory, most educated non-scientists felt humiliated by the technical details and despaired of ever understanding what Einstein had actually done. "Not even the old and much simpler Newtonian physics was comprehensible to the man in the street," lamented the *New York Times* in the wake of both relativity and quantum mechanics. "To understand the new physics is apparently given only to the highest flight of mathematicians."(3) Decades later, literary critic Lionel Trilling expressed the frustration of the humanist: "Physical science in our day lies beyond the intellectual grasp of most men.... This exclusion of most of us from the mode of thought which is habitually said to be the characteristic achievement of the modern age is bound to be experienced as a wound given to our intellectual self-esteem."(4)

The exclusion of non-specialists from technical science is not new. What is new is the profound sense of anxiety, alienation, and even resentment that existed across the divide separating what C. P. Snow called "the two cultures"—the sciences and the humanities.(5) Although quantum mechanics and the involvement of physical science in the century's wars and weapons of mass destruction contributed to that separation, the bond that had long joined science to culture began to erode almost immediately. Reflections of Einstein's seemingly unapproachable physics in the public mind, writes Yaron Ezrahi, "helped undermine science as a hallmark of public culture."(6)

By 1920 most physicists were incorporating the requirements and unifying aspects of special relativity into their work, while admiring general relativity from afar as the forefront of research moved away from relativity theory into atomic physics and quantum mechanics. But some also saw in relativity a chance to advance themselves and their careers by exploiting public anxieties. Nazi physicists Philipp Lenard and Johannes Stark provided the most extreme examples. The two Nobel laureates saw the confirmation of relativity theory as a threat to their status as experimentalists. Upon Hitler's rise to power in 1933, they gained professional advantage by formulating the Nazi ideology of "German" or "Aryan" physics. It was aimed, in their words, at freeing the German physics profession from so-called "Jewish physics," that is, modern theoretical physics—the "great dogmatic theories...Einstein's relativity theories, Heisenberg's matrix theory, Schrödinger's wave mechanics."(7) This would be accomplished by reemphasizing experimental research and, of course, by appointing themselves and other Aryans to positions once occupied by Jewish physicists. In a racist twist on relativism, science, they argued, was not objective, empirically testable, and universal, but a manifestation of the race (or ethnicity) of its practitioners.

Aryan physics and regime policies destroyed the leading status of the German physics profession. By driving some of the best physicists from Germany, among them Einstein, they contributed to the rise of physics in other nations, such as the United States, and in part to the failure of the war-time German atomic project to achieve even a working reactor. Einstein never set foot in Germany again.

Exploitation of public anxieties by physicists unfortunately occurred in practically every other industrialized nation. British physicists James Jeans and Arthur Eddington were among the most prominent and widely read popularizers of relativity theory in Britain and the United States. Their portrayal of the theory was wholly idealistic—even beyond what they themselves believed.(8) But it was a portrayal that the British and American public eagerly received for the reassurance it provided. After decades of debate over science and religion in the wake of evolution, Jeans and Eddington assured the Anglo-American public of relativity's limited human scope and of the wide latitude it, together with quantum indeterminism, supposedly allowed for free will, traditional beliefs, and individualistic democracy. Jeans conjectured in 1934 and again in 1942 that the ultimate understanding of nature will result in "the total disappearance of matter and mechanism, mind reigning supreme and alone." "The new physics," he wrote, "shows us a universe which looks as though it might conceivably form a suitable dwelling-place for free men, and not a mere shelter for brutes—a home in which it may at least be possible for us to mould events to our desires and live lives of endeavour and achievement."(9)

The success of idealistic interpretations in the West encouraged a backlash in the Soviet Union, where the dominant ideology was "dialectical materialism." As in other ideologies, science was regarded, not as an objective product of the universal search for truth, but—in antithesis to the Nazi insistence on race—as a manifestation of the economic relations of society.

Characteristically, both Marxist and Nazi ideologies, though so opposite in outlook, led their most dogmatic followers to react to relativity theory in much the same way: with suppression of the theory and persecution of its followers. During the 1920s, those eager for influence used the Western celebration of relativity as a triumph for idealism and individualism to stamp out the theory and its followers in the Soviet Union. Although others in the following decades, such as V. A. Fock, managed to rehabilitate the theory and themselves, they did so by introducing adaptations of the theory to its ideological context.(10)

As the meaning of relativity became blurred in the public mind under exploitative portrayals and critiques, philosophers began to take up the challenge of clarifying the content and implications of the new theory. Although only a few professional philosophers felt confident at first to attempt a philosophical analysis of the new physics, the eventual influence of the theory on twentieth-century philosophy was vast and far-reaching, gaining additional influence when joined with the issues raised by quantum mechanics. Of the many directions that philosophical research entered during the century, two of the most important entailed a clarification of the meanings of the fundamental concepts and of the nature of this remarkable theory itself.

One result of this research involved a preoccupation with time, reflected also within wider cultural circles. This entailed a shift, writes A. P. Ushenko, from the long-standing "metaphysics of material and mental things" in classical physics to a "metaphysics of events" stemming from relativity theory.(11) A stationary object, such as a train stopped on a track, is not just a thing, but, because of the time dimension, a thing moving through four-dimensional space-time—an event.(12) The similarities with non-metrical poetry and with Cubist art of the period, in which many temporal views of an object are displayed simultaneously, are striking—but Einstein vehemently denied any direct connection between the theory and the art.(13)

Another result of this research emerged in part as a defense of relativity theory through an appeal to positivism—the assertion that positive knowledge is founded only on empirical evidence. The new, relativistic meanings of concepts that had long been regarded as absolute seemed to entail numerous implications about the material world. Out of concern for the misapplication of these meanings to human affairs, there arose one of the prominent philosophical movements of the century, and one that gained particular allegiance in the United States and Great Britain—analytic and linguistic philosophy.

Following in the positivist tradition, Bertrand Russell and, especially, Ludwig Wittgenstein argued that only those statements that are empirically verifiable can have any meaning at all, while statements lying beyond empirical verification—specifically in the realms of theology, morals, and metaphysics—are, by definition, meaningless. This approach gained widespread influence in the hands of Moritz Schlick, Rudolf Carnap, and

the Vienna Circle of logical positivism. Schlick, a
Viennese professor, published one of the first philo-
sophical analyses of relativity theory in 1917. He had
come to philosophy from physics and maintained close
connections with Einstein and other leading physicists.
The circle eventually scattered to Britain and the United
States following the rise of Hitler and the annexation of
Austria.(14)

The influence of relativity theory on the Viennese is
discernible in two related elements of the circle's work:
Its insistence that the task of philosophy is to clarify the
meanings of concepts and statements, especially those in
science; and an admiration for physical science that
manifested itself in efforts to improve its methodology,
to unify its foundations, and to defend it philosophical-
ly from the criticisms of an anxious public.

Influenced by Russell and Wittgenstein, and inspired
by Einstein's definitions of concepts in terms of thought
experiments, the Vienna Circle argued that the only
meaningful scientific statements are ones defined by the
experimental operations with which they are used and
measured. This approach had much in common with
the "operationalism" espoused by American physicist
Percy W. Bridgman in his own attempt to make sense of

relativistic concepts.(15) "The final giving of meaning always takes place therefore through *deeds*," wrote Schlick. "It is these deeds or acts which constitute philosophical activity."(16)

The insistence on operations and empirical meaning was well suited for an assault on "metaphysics" and every non-empirical statement regarding human affairs, such as those offered by Oswald Spengler and the neo-romantics. For Spengler, the triumph of relativity signaled an ultimate "Decline of the West" for its supposed emphasis, unlike Eastern mysticism, on "lifeless" and "soulless" mechanistic materialism. Einstein's reiteration of classical determinism in relativity theory, and his rejection of quantum indeterminism, reinforced the neo-romantic critique of his work. Yet others, even Einstein, complained of the positivists' over-emphasis on empiricism and linguistic analysis. Einstein downplayed the positivist fear of metaphysics as a "malady."(17)

Another component of logical positivism was its concern for symbolic logic. Again drawing upon Wittgenstein and Russell, and enamored of Einstein's theory of relativity, the Viennese and others attempted to re-create the logical structure of scientific advance toward a new theory, such as special relativity. This was at first intended for descriptive purposes only, but as it began to prescribe how scientific advance does and should occur, it began to fall under the increasing skepticism of the emergent community of science historians. In addition, the sharp separation of theoretical (or analytical) statements from empirical statements, envisioned by the positivists, became increasingly untenable.

The history of science, rising to professional status following the Second World War, pointed out the disparity between science as it is actually practiced and the logical ideal promoted by positivist philosophers and some physicists, including, at times, Einstein. Thomas S. Kuhn, a leading critic of the logical positivist interpretation, even as it was later modified by Karl Popper, argued in his famous book, *The Structure of Scientific Revolutions* (1962), the need to include the contribution of non-logical, human, and social elements in any account of the course of scientific development. As a new and anxious age of nuclear cold war and the Vietnam era unfolded, many sought to modify the logical ideal of scientific research by taking into account social and cultural perspectives in order to achieve a more realistic portrayal of science as both a technical discipline and a historical phenomenon.

The historians' task required perspectives drawn from both the sciences and the humanities, but the separation of these two cultures had nearly approached the breaking point. The separation accelerated after the war amid fears of the results of scientific work, the political domination of science, the failure of schools and the media to educate the public about the many wonderful results of research, and the sometimes dogmatic and elitist attitudes of some of its practitioners.

Mutual animosity across the cultural divide was perhaps inevitable, but some recent writers have gone even further. The once revered light of reason is lately seen to have reached its limits, bringing a "twilight of the scientific age." In ways reminiscent of the relativist tradition, some have argued that scientific results are not logical or empirical statements at all but constructs created by social consensus and adapted to their social environment.(18) Still others, in ways equally reminiscent of earlier anxieties and efforts to gain social advantage at the expense of science, seek to eradicate some successful theories from the classroom by emphasizing their apparent conflict with theological tradition.

All of these responses to relativity theory, both positive and negative, over the century since Einstein's first relativity paper have influenced practically every form of cultural expression in an age of shifting values and meanings. At the same time the divide between science and culture, between scientists and the general public, has continued to grow. But Einstein himself saw the widening divide quite differently. His remarks, written in 1937, still reflect, as so often, a timeless ray of hope. "All religions, arts and sciences are branches of the same tree. All these aspirations are directed toward ennobling man's life, lifting it from the sphere of mere physical existence and leading the individual toward freedom."(19)

Notes

1. Einstein, "Epilogue: A Socratic Dialogue," in Max Planck, *Where is Science Going?* (New York: W. W. Norton, 1932), 201–221, on 209.

2. Einstein, "Principles of Research," 1918, in Einstein, *Ideas and Opinions,* Carl Seelig, ed. (New York: Modern Library, 1954/1994), 244–248; "What is the Theory of Relativity?" 1918, ibid., 248–253; "Physics and Reality," 1936, in Einstein, *Out of My Later Years* (New York: Philosophical Library, 1950), p. 96.

3. "A Mystic Universe," editorial, *The New York Times*, 28 July 1928, 14.

4. Lionel Trilling, *Mind in the Modern World: The 1972 Jefferson Lecture in Humanities* (New York: Viking Press, 1972), 13–14.

5. C. P. Snow, *The Two Cultures and the Scientific Revolution* (New York: Cambridge University Press, 1959).

6. Yaron Ezrahi, "Einstein and the Light of Reason," in *Albert Einstein: Historical and Cultural Perspectives, The Centennial Symposium in Jerusalem*, G. Holton and Y. Elkana, eds. (Princeton: Princeton University Press, 1982), 253–278, on 273.

7. Stark, *Nationalsozialismus und Wissenschaft* (Munich: Zentralverlag der NSDAP, 1934), 14. Also, Philipp Lenard, "Vorwort," to Lenard, *Deutsche Physik* (Munich: Lehmann's Verlag, 1936). See Alan Beyerchen, *Scientists under Hitler: Politics and the Physics Community in the Third Reich* (New Haven: Yale University Press, 1977), 123–140.

8. See Loren R. Graham, "The Reception of Einstein's Ideas: Two Examples from Contrasting Political Cultures," in Holton and Elkana, note 6, 107–136.

9. James Jeans, *Physics and Philosophy* (New York: Macmillan, 1942/1946), 307 and 216.

10. See Graham, note 8; Graham, *Science in Russia and the Soviet Union: A Short History* (New York: Cambridge University Press, 1993); Jeffrey L. Roberg, *Soviet Science under Control: the Struggle for Influence* (New York: Macmillan, 1998); Alexei Kojevnikov, *Stalin's Great Science: The Times and Adventures of Soviet Physicists* (London: Imperial College Press, 2004).

11. Andrew Paul Ushenko, "Einstein's Influence on Contemporary Philosophy," in *Albert Einstein: Philosopher-Scientist*, vol. 2. ed. P. A. Schilpp (LaSalle, IL: Open Court, 1949), 608.

12. See Bertrand Russell, *The ABC of Relativity* (New York: New American Library, 1925/1962), 139–140. The influence of temporal events on Einstein's formulation of his theory is explored by Peter Galison, *Einstein's Clocks, Poincaré's Maps: Empires of Time* (New York: W. W. Norton, 2003).

13. Letter from Einstein to Paul M. Laporte, 4 May 1946, in LaPorte, "Cubism and Relativity, with a Letter from Albert Einstein," *Art Journal*, 25:3 (1966), 246. For more on influences on art and literature, see Gerald Holton, "Einstein's Influence on the Culture of Our Time," in

Holton, *Einstein, History, and Other Passions: The Rebellion Against Science at the End of the Twentieth Century* (Reading, MA: Addison-Wesley, 1996), 125–145.

14. An early history: Viktor Kraft, *The Vienna Circle: The Origin of Neo-Positivism, A Chapter in the History of Recent Philosophy*, trans. Arthur Pap (New York: Philosophical Library, 1953). The anti-metaphysical approach is proclaimed by Carnap, "The Elimination of Metaphysics through Logical Analysis of Language," in *Logical Positivism*, A. J. Ayer, ed. and trans. (New York: Free Press, 1959).

15. Percy E. Bridgman, *The Logic of Modern Physics* (New York: Macmillan, 1927). See Holton, note 13, 221–227.

16. Schlick, "The Turning Point in Philosophy," in Ayer, note 14, 57, his italics.

17. Einstein, "Remarks on Bertrand Russell's Theory of Knowledge," 1944, in *Ideas and Opinions*, note 2, 8–12, on 12.

18. For instance, Paul Horgan, *The End of Science: Facing the Limits of Knowledge in the Twilight of the Scientific Age* (Reading, MA: Addison-Wesley, 1996); Paul Feyerabend, *Against Method* (London and New York: New Left Books, 1975); Stephen Shapin, *A Social History of Truth: Civility and Science and Seventeenth Century England* (Chicago: University of Chicago Press, 1994).

19. Einstein, "Moral Decay," in Einstein, *Out of My Later Years*, note 2, 9.

SELECTED BIBLIOGRAPHY

Baierlein, Ralph. *Newton to Einstein: The Trail of Light, An Excursion to the Wave-Particle Duality and the Special Theory of Relativity*. New York: Cambridge University Press, 2001.

Bernstein, Jeremy. *Albert Einstein and the Frontiers of Physics*. New York: Oxford University Press, 1996.

Bodanis, David. *E=mc²: A Biography of the World's Most Famous Equation*. New York: Walker, 2000.

Born, Max. *Einstein's Theory of Relativity*. Revised edition. New York: Dover, 1965.

Cassidy, David C. *Einstein and Our World*. 2nd edition. Amherst, NY: Humanity Books, 2004.

Einstein, Albert. *The Collected Papers of Albert Einstein*. Multiple volumes. John Stachel et al., eds. Princeton: Princeton University Press, 1987–. With parallel English translation in separate series.

————. *The Meaning of Relativity*. Princeton, NJ: Princeton University Press, 2005.

Einstein, Albert and Leopold Infeld. *The Evolution of Physics: The Growth of Ideas From Early Concepts to Relativity and Quanta*. New York: Free Press, 1967.

Einstein, A., H. A. Lorentz, H. Minkowski, and H. Weyl. *The Principle of Relativity: A Collection of Original Papers on the Special and General Theory of Relativity*. New York: Dover, 1924.

Feynman, Richard P. *Six Not-So-Easy Pieces: Einstein's Relativity, Symmetry, and Space-Time*. Reading, MA: Addison-Wesley, 1997.

Galison, Peter. *Einstein's Clocks, Poincaré's Maps: Empires of Time*. New York: Norton, 2003.

Gardner, Martin. *The Relativity Explosion*. New York: Vintage, 1976.

Geroch, Robert. *General Relativity From A to B*. Chicago: University of Chicago Press, 1978.

Hartle, James B. *Gravity: An Introduction to Einstein's General Relativity*. San Francisco: Addison-Wesley, 2003.

Hawking, S. W. and G. F. R. Ellis. *The Large-Scale Structure of Space-Time*. Cambridge: Cambridge University Press, 1973.

Holton, Gerald. *Einstein, History, and other Passions*. Cambridge, MA: Harvard University Press, 2000.

————. *Thematic Origins of Scientific Thought: Kepler to Einstein*. Cambridge, MA: Harvard University Press, 1988.

Miller, Arthur I. *Einstein, Picasso: Space, Time, and the Beauty that Causes Havoc*. New York: Basic Books, 2001.

Misner, Charles W., Kip S. Thorne, and John Archibald Wheeler. *Gravitation*. San Francisco: Freeman, 1973.

Pais, Abraham. *"Subtle is the Lord...": The Science and the Life of Albert Einstein*. New York: Oxford University Press, 1982.

Penrose, Roger. *The Road to Reality: A Complete Guide to the Laws of the Universe*. New York: Knopf, 2005.

Rindler, Wolfgang. *Relativity: Special, General, and Cosmological*. New York: Oxford University Press, 2001.

Stachel, John. *Einstein from "B" to "Z"*. Boston: Birkhäuser, 2002.

———, ed. *Einstein's Miraculous Year: Five Papers That Changed the Face of Physics*. Princeton, NJ: Princeton University Press, 1998.

Wald, Robert M. *General Relativity*. Chicago: University of Chicago Press, 1984.

INDEX